AN INTRODUCTION TO BUILDING SERVICES

Christopher A. Howard

MPhil, DipQS, FRICS

MACMILLAN
EDUCATION

First published 1988

Published by
MACMILLAN EDUCATION LTD
Houndmills, Basingstoke, Hampshire RG21 2XS
and London
Companies and representatives
throughout the world

Printed in Hong Kong

British Library Cataloguing in Publication Data
Howard, Christopher A.
   An Introduction to building services. —
   (Macmillan building and surveying).
   1. Buildings — Environmental engineering
   I. Title
   696   TH6021

ISBN 0–333–43462–5

Series Standing Order

If you would like to receive future titles in this series as they are
published, you can make use of our standing order facility. To place
a standing order please contact your bookseller or, in case of difficulty,
write to us at the address below with your name and address and the
name of the series. Please state with which title you wish to begin your
standing order. (If you live outside the United Kingdom we may not
have the rights for your area, in which case we will forward your order
to the publisher concerned.)

Customer Services Department, Macmillan Distribution Ltd
Houndmills, Basingstoke, Hampshire, RG21 2XS, England.

AN INTRODUCTION TO BUILDING SERVICES

*To my family
for their patience and
understanding throughout the production
of this work*

CONTENTS

PREFACE

This book applies a principles approach throughout and is designed to meet the needs of those whose responsibility is the administration, control, and inspection of buildings. The parameters on which the text is based include an appreciation of the choice options available, identification of major components, and an insight into operation.

Many in authoritative positions of management require a broad understanding of services but do not need the detailed design knowledge which would be required by an engineer. Survey inspections for maintenance, repair and safety may also require this approach.

To complement each section within the text, many references are made to appropriate British Standards, Codes of Practice and the *Building Regulations 1985*.

Christopher A. Howard

ACKNOWLEDGEMENTS

The author acknowledges the willingness of many manufacturers within the building industry to permit the reproduction of text illustrations of their products.

Special thanks are given to Elizabeth and Paul Hodgkinson for the excellence of their artwork.

LIST OF FIGURES

1 COLD WATER INSTALLATIONS — SUPPLY AND DISTRIBUTION

INTRODUCTION

The extent of rainfall in the UK should provide a more than adequate supply of water to meet the needs of residential users. Unfortunately, some of the total stock of water generated by rainfall will be polluted whether it collects in water courses such as rivers or streams, or in natural water accumulations such as lakes and underground deposits. Pollution by man rather than natural pollution is the major criteria in considering the extent of the available supply.

Whether the supply is naturally or unnaturally polluted, treatment processes will ensure the purity and palatability for the consumer. The amount of treatment necessary will often relate to the sources of supply. As a generalisation, water drawn from deep underground sources tends to be least polluted and most palatable (although it may be the hardest), while that which is obtained from shallow wells or rivers in developed areas tends to be the most polluted and least palatable.

Hardness of water is a condition which may be temporary or permanent. It arises in water because of the absorption of chemicals into the supply as the water passes through the ground. The chemicals causing hardness are calcium and magnesium sulphates and bicarbonates of calcium or magnesium. Temporary hardness due to bicarbonates can be removed by boiling, while boiling will not affect permanent hardness due to the presence of sulphates. For the treatment of large volume supplies it would not be practicable to apply boiling and Clark's process of lime water treatment is generally used.

When examining the range of available water treatment processes, a division of applicable processes will arise from the quantity of purified supply demanded. Consequently, different treatments will be applied to the isolated individual consumer than to the consumer in densely populated areas. Isolated supplies drawn from water courses for the individual consumer are often subjected to only very basic screening and filtration, while by contrast far more extensive treatment will occur for the large volume supplies.

Large-scale treatment will ensure the removal of suspended organic and inorganic material, and dissolved organic impurities. One of the main considerations will be the destruction of pathogenic bacteria which may be the vehicle for the transmission of certain diseases (typhoid and cholera). The treatment of large-scale supplies broadly involves the following:

(a) straining to remove larger suspended or floating objects;
(b) sedimentation to remove smaller suspended impurities in particle form;
(c) aeration to inject oxygen and reduce the carbon dioxide content;
(d) filtration to remove all remaining suspended particles and reduce the bacterial content;
(e) sterilisation (usually chlorine based) for the destruction of bacteria.

The provision of a cold water supply to a building is often to satisfy three demands: water for drinking, water for general usage, and water for fire fighting.

In the United Kingdom it is a statutory requirement of the Water Acts[1] that every dwelling shall be supplied with drinking water. Consumption for general usage includes water for sanitation and laundry purposes, while that needed for fire fighting may considerably influence the size of the main routed to the premises.

Supplying cold water may be best considered by examination of the geometry of the buildings served, since large and tall buildings present numerous installation and design problems which are not found in smaller low rise buildings.[2]

SUPPLIES TO LOW RISE DWELLINGS

Domestic low rise dwellings are generally of single, two, or three storey construction. Individual premises are connected to the water main by a service pipe branch as shown in figure 1.1.

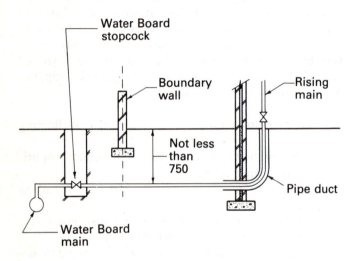

Figure 1.1   Cold water supply — the service pipe

The features of the service pipe include:

(i) the facility for the water authority to disconnect the supply by use of a stopcock located outside the house boundary;
(ii) laying of the pipe at a depth which will prevent freezing; and
(iii) the provision of a duct to carry the supply into the building through the wall and floor.

In older property the water service pipe is often in lead, whereas more modern premises are provided with plastic pipes, such as 'Vulcathene'. The material used for the service pipe often has significance to the electrical installation which has traditionally been earthed on to this buried pipe and consequently the use of plastic pipes will neces-

sitate a separate earthing facility. Attempts to earth the electrical installation on to plastic service pipes is an incorrect detail occasionally found by the surveyor during survey inspections.

Once the service pipe emerges from the floor of the premises it is referred to as the rising main, and it is generally at the point of entry through the floor that a stopcock[3] is provided which allows the resident to close down the incoming supply. Figure 1.2 shows a typical cold water installation for a two storey dwelling.

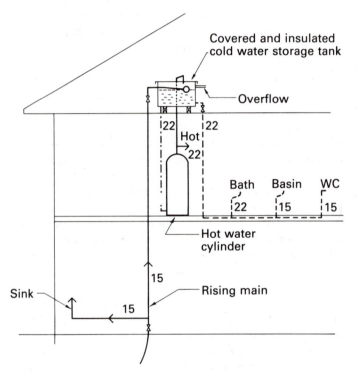

Figure 1.2   Cold water installation — two storey domestic

Mains-pressure water is carried by the rising main to the cold water storage tank where water is stored for general-purpose use. Drinking water is taken in a branch from the rising main to the kitchen sink.

A mains-pressure branch may also be taken from the rising main to a ground floor WC, and if this is the case it will be necessary to provide a high-pressure ball valve assembly to the WC cistern. The ball valve is used to maintain the water level in cold water storage tanks; its typical component parts are shown in figure 1.3

The ball valve represents probably the most problematic component of a cold water system since it requires regular maintenance and periodic replacement. A stopcock is positioned on the rising main at the cold water storage tank to allow a plumber to close down the supply while attending to this valve.

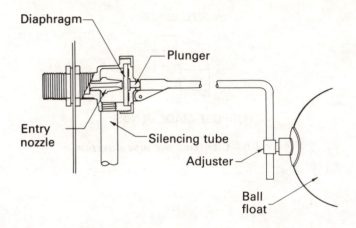

Figure 1.3   A typical ball-valve assembly

General-purpose water is supplied from the cold water storage tank to the hot water cylinder and the bathroom sanitary appliances. This water is now at gravitational (tank) pressure rather than mains pressure. Note the provision for stopcocks. An alternative to the stopcock for the restriction of flow is the gate valve; figure 1.4 illustrates the difference internally between these control valves. Externally they can be identified by the crutch head of the stopcock and the wheel head of the gate valve.

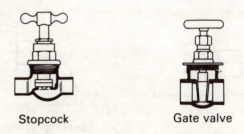

Figure 1.4   Stopcocks and gate valves

The stopcock is satisfactory for controlling the flow of supplies of cold water at local mains pressure where distortion in the bore of the flow is not a critical factor. By contrast, the gate valve, which works on a sluice gate principle, may be suited to higher-pressure flows while also having the advantage of allowing near complete bore flow. The latter point is of considerable importance in hot water conveyance to minimise pressurisation losses during water circulation.

Cold water storage tanks located in the roofspace should be covered and insulated against freezing. This is of particular importance since current Building Regulations require the roofspace to be ventilated to reduce the possibilities of condensation.[5] The capacity of the storage vessel should be sufficient to provide a small reserve of general-purpose water in the event of a temporary discontinuation of mains supply, and tanks[6] for domestic installations are generally around 182 litres (40 gallons) capacity.

Lead piping has been used extensively during the first half of this century for cold water systems but today copper is the most popular material (BS 2871, Tables X, Y and Z[7]). Copper pipes are specified according to outside diameter:

		outside diameter (mm)
$\frac{1}{2}''$	=	15
$\frac{3}{4}''$	=	22
$1''$	=	28
$1\frac{1}{4}''$	=	35
$1\frac{1}{2}''$	=	42
$2''$	=	54

Pipe jointing in copper may be achieved by the use of capillary fittings (soldered) or compression fittings (nutted), see figure 1.5. The latter are considerably more expensive but have the advantage that they are easy to dismantle.

Other pipe materials for cold water installations include stainless steel, polythene, acrilonitrile–butadiene–styrene (ABS) and unplasticised polyvinyl chloride (uPVC).[8]

The choice of pipe material may be made in terms of cost or performance. With respect to the latter point, some observations of the alternative materials may be listed:

Capillary elbow   Compression elbow   Bending spring   Bending machine

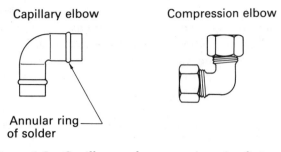

Annular ring
of solder

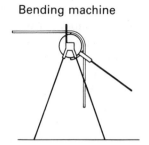

Figure 1.5   Capillary and compression pipe fittings

copper   — may be more easily handled than
                other metals (such as steel)
                — is versatile in that it may be used for
                cold water, hot water and gas

stainless steel — provides for good standards of
                hygiene which may make it particu-
                larly suitable for food processing
                and hospitals
                — provides a good appearance

ABS   — has high impact strength and non-
                toxicity which make it ideal for the
                chemical and pharmaceutical indus-
                tries as well as for domestic installa-
                tions

uPVC   — the impact strength of this material
                deteriorates rapidly at low tempera-
                tures

Another factor considered when selecting a pipe material for cold water installations will be the suitability of
the material for the hot water installations. In domestic
work there would be a natural resistance to using two
different pipe materials, one for cold water and one for
hot. This point may limit the application of the plastic
pipes listed since the maximum working temperature is
80°C for ABS and 60°C for PVC.

When making changes in direction in pipework either
purpose-made fittings or 'made bends' may be employed.
Figure 1.6 shows the use of a bending spring for smaller-
diameter thin-walled pipes, a bending machine for
thicker-walled pipes and a purpose-made bend, all of
which may be used to effect a 90° change in direction.

Purpose-made fittings are also used to make junctions
between pipes or to connect pipes to appliances and
equipment. Some commonly used fittings are shown in
figure 1.7.

IN-SITU BENDS

PURPOSE-MADE BENDS

Figure 1.6   Changing the pipe direction

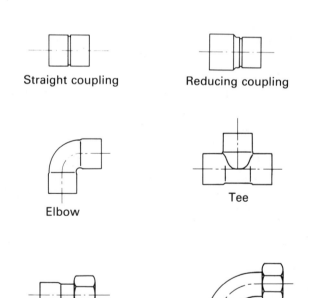

Straight coupling   Reducing coupling

Elbow   Tee

Straight tap connector

Bent tap connector

Figure 1.7   Capillary pipe fittings

SUPPLIES TO HIGH RISE BUILDINGS

As the supply pressure of the Water Board main is unlikely to allow direct delivery to the upper floors of tall buildings, assistance has to be given to the main to enable it to reach these levels.

The design of cold water installations to high rise buildings is influenced by the fluctuations in mains pressure caused by peak demand and by the structural imposition imposed by the storage of water (1 litre = 1 kg). The dead weight of water may necessitate the zoning of stored supplies if loads are not to be concentrated at one level.

Buildings in multiple occupancy may require significant volumes of water for general-purpose use and, as this is supplied to the user at gravitational pressure, storage has to be above the point of delivery. With multi-storey blocks of flats, individual cold water storage tanks may be provided to each residence or communal storage at roof level may be the alternative. In tall buildings the main will be allowed to serve drinking water to the floors within its capacity and pumps will directly or indirectly assist in conveying the water to the upper levels. To have pumps continually in operation would be inefficient and present a considerable maintenance commitment, and consequently techniques are employed to prevent continual pump use.

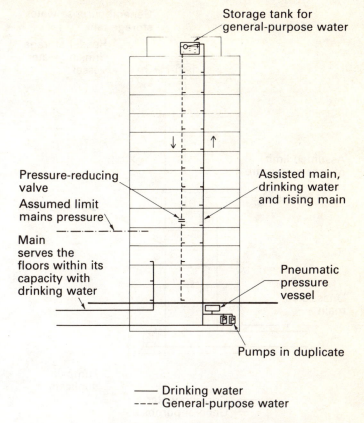

Figure 1.8   The pneumatic system of cold water delivery

The Pneumatic System

The pneumatic system of cold water delivery is illustrated in figure 1.8. This system uses compressed air within a pressure vessel to assist the main.

The pump forces the water into the pressure vessel which contains a volume of air. On reaching the desired air pressure level the pumping ceases and the compressed air now provides the driving force behind the mains water. As water is used by the occupants of the building, the air in the pressure vessel expands and the pressure falls. At a pre-determined low level of pressure, the pump is again brought into operation. In the event of air being absorbed into the water during pressurisation, a small compressor is on hand to replenish the air supply.

The system illustrated may be used for a block of around 15 storeys, with general-purpose water stored at roof level to serve all units.

The Header Storage System

The name of this system arises from the enlargement of the rising main which occurs just before connection to the roof-level cold water storage tanks. By enlarging the rising main into a cylinder above the last user outlet, a storage vessel is formed which could supply the upper levels of the premises with gravity-pressure drinking water if the mains supply were to be interrupted. The capacity of the header storage cylinder is generally 5–7 litres per dwelling served.

In this system the pump is activated either by the operation of a float switch located in the general-purpose cold water storage tank or by a similar switch inside the header cylinder. Figure 1.9 illustrates a typical system.

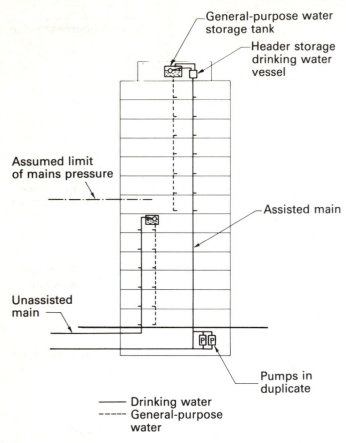

Figure 1.9   The header storage system

Enclosed Tank Systems

Traditionally cold water for drinking purposes is taken directly from the mains at mains pressure, since storage can lead to possible contamination or loss of palatability. However, provided an enclosed, insulated and ventilated tank is used, storage of drinking water is permissible.

There are two common arrangements of system which provide for the storage of drinking water.

Figure 1.10 illustrates one system which employs enclosed hygienic tanks for drinking water storage either from:

(a) one large drinking water tank to serve all outlets from roof level; or

(b) a smaller tank provided to serve a number of floors in a zoning arrangement; or

(c) individual tanks to each user.

In this system the general-purpose water is stored in separate storage tanks following the arrangement in either (a), (b) or (c) above.

Figure 1.11 shows an alternative system where a large capacity tank located at ground or basement level is used to supply both drinking water and general-purpose water. From this tank the water is pump-assisted to roof level, where it feeds a tank for gravity-supply drinking water and another for gravity-supply general-purpose water.

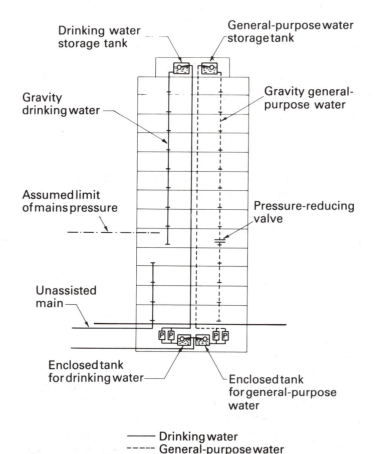

Figure 1.10   An enclosed tank system

Pipe Sizing[9]

The size of a cold water rising main will be influenced by the number of connections made to it, the frequency with which those connections draw water, and the mains pressure available to the premises.

In domestic installations a 15 mm rising main will suffice in individual low rise buildings but, where multistorey buildings are served, significantly larger diameters will apply. The diameters of gravity-pressure water supplies will similarly depend on the number of outlets served, their capacity, the frequency with which they are used and the pressure of delivery.

Gravity supplies of cold general-purpose water in domestic plumbing will be taken to the hot water cylinder and bathroom sanitary appliances. It should be remembered that the diameter of the supply to the bath is generally 22 mm, while the washbasin and WC take a 15 mm supply.

SANITARY APPLIANCES[10]

Many sanitary appliances are readily recognisable, but variations of types may need to be known.

Figure 1.12 shows variations of types of washbasins[11] (also called handbasins or lavatory basins). The types illustrated are the pedestal, the bracket and leg, and the cantilever. An illustration of the term *range* is also given, this being a description which applies to three or more basins in line. The Belfast sink is included to show the incorporation of an integral overflow into an appliance.

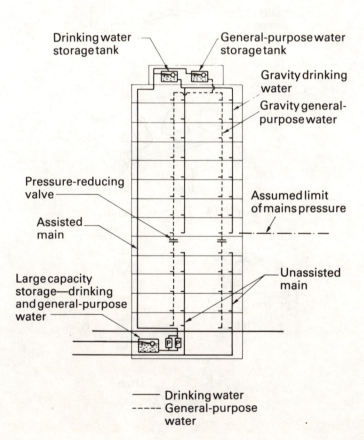

Figure 1.11   An enclosed tank system

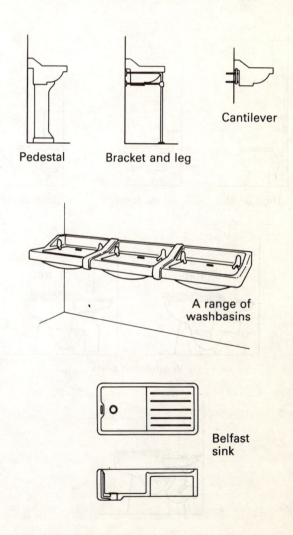

Figure 1.12   Sanitary appliances — washbasins and the Belfast sink

Figure 1.13 shows variations in the types of WC available;[12] these include high-level, low-level and closed coupled details, and also P-trap and S-trap washdown pans. Where closed coupled WC details are used, these are often provided with a siphonic pan as the low flush pressure requires suction assistance to clear the bowl.

Figure 1.14 shows the difference between stall, slab and bowl urinals,[13] while figure 1.15 shows a variety of taps for appliances.

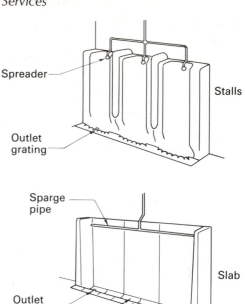

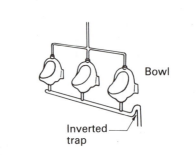

Figure 1.14   Types of urinal

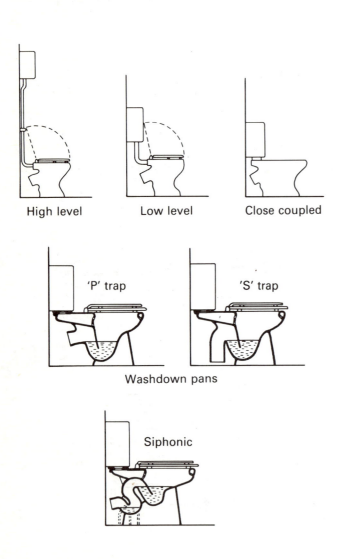

Figure 1.13   Types of WC

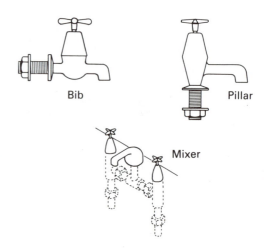

Figure 1.15   Bib, pillar and mixer taps[14]

REFERENCES

1 Water Acts 1956, 1968.
2 Ministry of Housing and Local Government, *Design Bulletin 3, Part 6: 1965 Services cores in high rise flats. Cold water services.*
3 *BS 1010: Part 2: 1973 Draw off taps and above ground stopvalves; BS 5433: 1976 Specification for underground stopvalves for water services.*
4 *BS 1968: 1953 Floats for ballvalves (copper); BS 2456: 1973 Floats (plastic) for ballvalves for hot and cold water.*
5 *Building Regulations 1985.* Part F, Approved Document F, clauses 1.4 and 1.5, p.6
6 *BS 417: Part 1: 1964, Part 2: 1973 Galvanised mild steel cisterns and covers, tanks and cylinders.*
7 *BS 2871: Part 1: 1971 Copper tubes for water, gas and sanitation.*
8 *BS 4127: Part 2: 1972 (1981) Light gauge stainless steel tubes; BS 1972: 1967 Polythene pipes (Type 32) for above ground use for cold water services; BS 5391: Part 1: 1976 Specification for acrylonitrile–butadiene–styrene (ABS) pressure pipes; BS 3505: 1968 (1982) Unplasticised PVC pipes for cold water services.*
9 *CP 310: 1965 Water supply,* Appendix.
10 *BS 6465: Part 1: 1984 Code of practice for scale of provision, selection and installation of sanitary appliances.*
11 *BS 1188: 1974 Ceramic washbasins and pedestals; BS 5506: 1977 Specifications for wash basins, Parts 1, 2 and 3.*
12 *BS 5503: 1976 Specification for vitreous china washdown WC pans with horizontal outlet; BS 5504: 1977 Wall hung WC pans.*
13 *BS 4880: Part 1: 1973 Stainless steel slab urinals; BS 5520: 1977 Specification for vitreous china bowl urinals.*
14 *BS 5412: 1976 and BS 5413: 1976 Specification for the performance of draw-off taps.*

2 HOT WATER GENERATION AND DISTRIBUTION

INTRODUCTION

Both the type of premises served with hot water and the nature of the fuel used will have an influence on the method employed for hot water generation.[1] The methods of generation may be listed as follows:

(a)   by using electric immersion heaters;[2]
(b)   by using boilers — gas,[3] oil, or solid fuel fired;
(c)   by using gas heaters;[4]
(d)   by using solar heaters.[5]

HOT WATER GENERATION BY ELECTRIC IMMERSION HEATERS

The generation of hot water by electric immersion heaters may be considered to be one of the simplest forms of generation. Heating elements are usually attached to the top of the hot water cylinder which is well insulated to help improve the efficiency of this method (figure 2.1).

Thermostats are used to control the temperature of the stored hot water, and time switches are commonly employed for pre-set activation of the heater.

As electricity is often regarded as an expensive method of hot water generation, a variable demand immersion heater may be seen as a more economic means of meeting the pattern of hot water usage. Such arrangements are achieved by the use of two heating elements in the hot water cylinder: a small element (perhaps 0.5 kW) at the top which can supply a limited amount of hot water for washing up and the like, and a larger element (perhaps 2 kW) towards the bottom which can heat the whole cylinder and cater with the demands made by bathing. Figure 2.2 shows a typical variable demand cylinder.

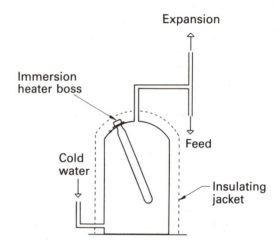

Figure 2.1   A domestic electric immersion heater

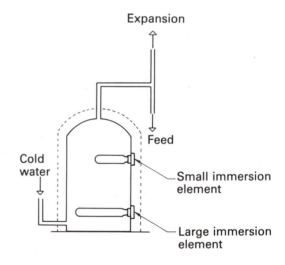

Figure 2.2   A variable demand electric immersion cylinder

To improve the ease with which residential properties are provided with a means of hot water generation and cold water storage, the use of composite plumbing units has emerged. Figure 2.3 shows a boxed composite unit containing an electrically heated hot water cylinder and a cold water storage tank.

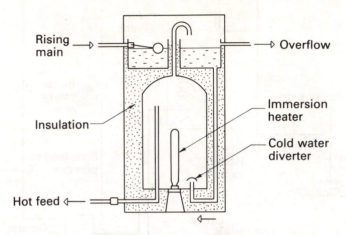

Figure 2.3   Composite plumbing packages — electric hot water heating and cold water storage

An additional form of electric immersion heating is provided by the instantaneous heater. Shower water heaters are of this type, these taking mains-pressure cold water supplies.

HOT WATER GENERATION USING BOILER SYSTEMS

When boilers are used for the generation of hot water they are generally also feeding a space heating circuit. In the notes that follow the ability to provide space heating is ignored, concentration being on hot water generation. Boiler hot water systems may be divided into two arrangements: direct and indirect systems.

Direct systems are typified by the open-fire back boiler arrangements used extensively in residential property built before the Second World War — see figure 2.4.

The disadvantage of the direct system may be seen in the free flow of water from the primary pipework into the hot water cylinder, with the injection of fresh cold water into the cylinder and primary piping each time hot water is drawn off. In hard water areas this would encourage pipe

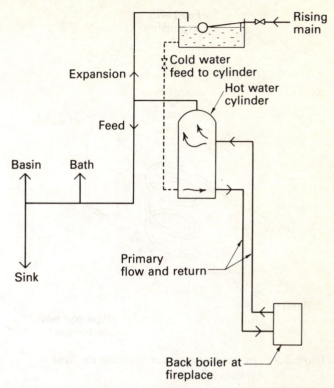

Figure 2.4   Direct hot water generation by open-fire back boiler

furring and consequently restrict flow, thus decreasing the efficiency.

Modern boiler generation arrangements, in contrast, are usually indirect systems in which the primary heat generation pipework becomes an enclosed circuit. This arises through the use of a heat exchanger which acts as the heating element inside the hot water cylinder. Hot water circulates through the heat exchanger in the same way as it would through a central heating radiator. The flow from the boiler through the heat exchanger may be by gravity under thermal circulation or alternatively the hot water may be pumped. The use of a pumped primary is normally associated with a layout where there is a considerable distance between the boiler unit and the hot water cylinder.

Heat exchangers may be simply a pipe coil (figure 2.5), or an annular cylinder profile.

The primary pipework, heat exchanger and boiler will have to be filled with water initially and topped up as necessary by the provision of a water supply solely for the use of that circuit. The cold water tank used for this purpose may be much smaller than that which is needed to

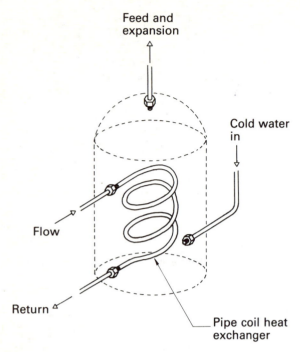

*Figure 2.5   An indirect cylinder showing the heat
              exchanger[6]*

store general-purpose water for the premises. An 18 litre
(4 gallon) feed and expansion tank will generally fulfil this
role, while the common size of the cold water storage tank
is typically 182 litres (40 gallons). An example of an
indirect system for hot water generation is shown in figure
2.6.

The generation of hot water on a communal rather than
on an individual basis may be provided by group or district
heating schemes where large capacity boilers are remotely
sited to provide heat generation. Such systems are considered in chapter 5.

HOT WATER GENERATION BY GAS HEATERS[3]

Apart from the use of the gas central heating boiler, other
methods of gas fired hot water generation include:

(a)   gas circulators;
(b)   gas single and multiple-point instantaneous heaters;
(c)   gas storage water heaters.

Gas circulators are effectively miniature boilers which
are connected to the hot water storage cylinder by flow

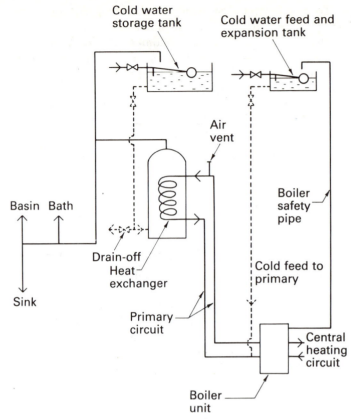

Figure 2.6   Indirect hot water generation

and return pipework in a similar fashion as in the indirect
method just discussed. As with the central heating boilers,
such circulators may have a pipe flue or balanced flue for
the expulsion of the combustion wastes (see chapter 8).

Instantaneous heaters may be located directly over the
sink (single-point) or may be remotely located to serve all
the appliances with hot water (multi-point). Both are
connected to the incoming mains water supply and have
burners which ignite as the tap to an appliance is opened.
Direct delivery to the sanitary appliance means that the
need for a hot water cylinder of the conventional type is
eliminated.

Gas storage water heaters are forms of insulated container which consist of a storage tank with an integral
water heater. The burners are activated when water
drawn from an appliance causes the temperature of the
stored water to fall below a pre-set level. Again this
system eliminates the need for a conventional hot water
storage cylinder.

SOLAR GENERATION OF HOT WATER

The use of solar energy for space heating is included in chapter 5. Solar collectors for hot water generation are generally of the same basic construction as those used to provide water for space heating. Both types rely on the 'greenhouse effect' of short-wave solar rays penetrating the glass or plastic cover of the collector. When penetration occurs, the short-wave radiation is converted to long wave which is less easily released through the glass cover, and the result is a temperature rise inside the collector.

During reasonable sunlight, water pumped through solar collectors can reach temperatures of around 50°C. A pre-determined temperature difference between the stored heated water and the solar collector deems that it is worthwhile activating the pumps to allow circulation (figure 2.7).

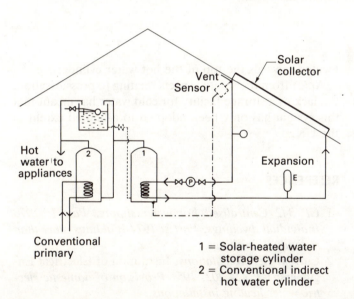

Figure 2.7   Solar generation of hot water

In this arrangement the solar collector simply replaces the boiler of the more conventional hot water system. The solar collector heats the water within the solar indirect cylinder number 1, which in turn provides a warm rather than a cold water supply to cylinder number 2, which feeds the sanitary appliances.

PRIMATIC INDIRECT HOT WATER CYLINDERS

Because the indirect method of hot water generation uses an enclosed primary circuit, this necessitates the use of the small feed and expansion cold water tank for filling the primary as already described. This cold water tank requires connection to the rising main and often substantial quantities of piping and fittings to connect it to the primary circuit. On major housing refurbishment schemes the conversion from direct hot water generation to indirect generation is common. If a method of filling the primary circuit of the indirect system could be found which eliminated the need for the feed and expansion tank and its associated pipework, the cumulative savings could be considerable. One attempt at such savings was made through the adoption of primatic indirect cylinders (figure 2.8).

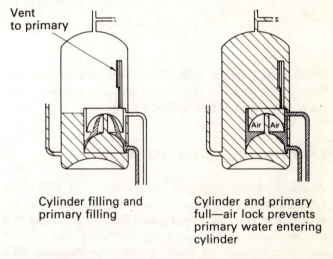

Figure 2.8   Primatic hot water cylinders

The principle of this cylinder relates to the special nature of its heat exchanger. The heat exchanger consists of two chambers between which there will be an air lock when the system is filled. This arrangement allows the primary circuit to be filled by water taken from within the hot water cylinder. However, when the system is filled and the boiler commissioned, water flowing in the flow and return of the primary should not be allowed to mix with the body of water inside the hot water cylinder as a result of the air lock.

In principle, the idea of the primatic cylinder means a considerable saving on the costs of the indirect system.

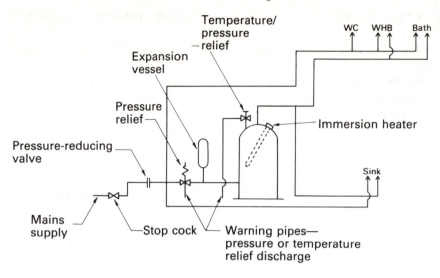

Figure 2.9   The unvented hot water system

Unfortunately some difficulties have been experienced with the air locks of this detail, with a result that it has been only adopted to a limited extent.

UNVENTED HOT WATER SYSTEMS[7]

A recent development in the provision of hot water supplies for domestic and other users has been the unvented hot water system. Figures 2.4 and 2.6 illustrated the provision of hot water by the direct and indirect methods. In both of these systems the outgoing pipe carrying hot water from the hot water storage cylinder is routed upwards to terminate over the cold water storage tank, thereby providing a means by which any adverse pressure within the hot water cylinder can be relieved. As a consequence, this length of pipe is referred to as the 'feed and expansion pipe'.

In contrast to the direct and indirect methods, the unvented layout, which has been extensively adopted outside the UK, has no ventilating facility on the hot water feed from the storage cylinder (figure 2.9).

To accommodate the pressures that may occur within the system, there is the facility for relief on the incoming mains cold water supply in the form of a ventilation valve and expansion vessel. A pressure relief valve is also located towards the top of the hot water cylinder.

Apart from the safety aspects relating to pressurisation, the lack of a storage facility for cold water has meant that this system has only been adopted to a limited extent in the UK.

REFERENCES

1 *CP 342 Centralised hot water supply. Part 1: 1970 Individual dwellings, Part 2: 1974 Buildings other than individual dwellings.*
2 *Current IEE Regulations*, Institution of Electrical Engineers; *CP 324.202: 1951 Provision of domestic electric water heating installations.*
3 *BS 5258: Part 1: 1975 Safety of domestic gas appliances; BS 6332: Part 1: 1983 Specification for thermal performance of central heating boilers and circulators.*
4 *BS 5546: 1979 Code of practice for installation of gas hot water supplies for domestic premises.*
5 *BS 5918: 1980 Code of practice for solar heating systems for domestic hot water.*
6 *BS 853: 1981 Specification for calorifiers and storage vessels for central heating and hot water supply.*
7 *BS 6283: 1982 Safety devices for use in hot water systems.*

3 DRAINAGE ABOVE GROUND — DISPOSAL INSTALLATION PIPEWORK

INTRODUCTION

This chapter concentrates on the means by which sanitary appliances are connected to the below-ground drainage system. For efficient working of a disposal installation pipework system, a number of design criteria should be fulfilled.[1] This applies to the smallest domestic system but becomes particularly important when related to large communal installations. By careful consideration of the design criteria, the creation of adverse conditions should be prevented.

To this end, the capacity of the appliances served by the installation and the frequency with which they are used are of paramount importance.[2] Certain appliances cause a large (although momentary) discharge and this tends to be reflected in the diameter of disposal pipe to which they are connected. Flow rates are:

WC   — in excess of 2 litres per second
bath   — about 1 litre per second
washbasin — about 0.5 litres per second

The diameter of the common pipes (stacks) is of particular concern to the designer as allowance must be made for the cumulative effect of all the discharges to be carried. To aid the selection of the correct stack diameter, a theoretical weighting of Discharge Unit Values has been developed. This weighting system reflects the appliance discharge capacity and the likely frequency of use. Some typical values for hotel premises are shown below:

Appliance	Frequency of use (minutes)	Discharge Unit Value
WC (9 litre)	20	7
Bath	30	18
Washbasin	10	3

By adding the Discharge Unit Values for all of the appliances connected to a stack, a suitable stack diameter may be found by reference to another table.

FLOW PATTERN AND AIR PRESSURE FLUCTUATIONS

When reference is made to pipes conveying discharges from sanitary appliances, two classifications of pipe are used:

Waste pipes — conveying water-bound discharges from sinks, baths, showers, washbasins and bidets
Soil pipes   — conveying human discharges from WCs or urinals

If we examine the way discharges move in these pipes, the pattern of flow may be readily visualised in near horizontal pipes. However, the flow pattern within vertical pipes is considerably different, as shown in figure 3.1.

At a certain capacity of flow, both pipes of course could be filled and, in such a situation, compression of the air in front of the flow would be the result. Even below full capacity, the bore of the pipe may be momentarily filled by the creation of either a 'jump' or a 'plug' formation (figure 3.2).

The primary function of the trap is to effect a seal against foul air in the pipework system,[3] thus preventing its entry into the building. Liquid seal depths to achieve this are generally 75 mm on pipes of up to 50 mm diameter, and 50 mm for larger pipes. The range of traps includes:

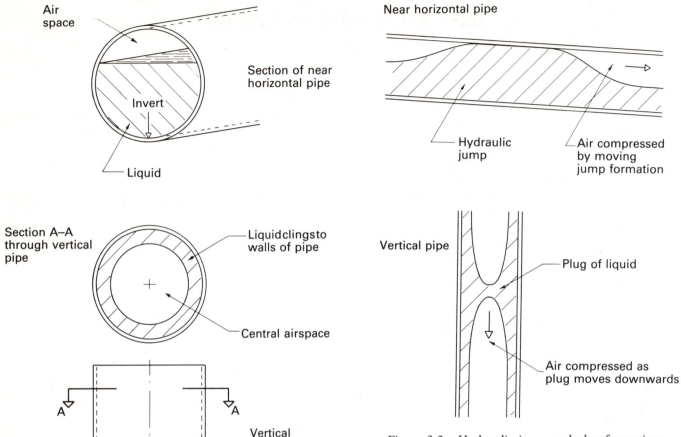

Figure 3.1   Flow patterns in near horizontal and vertical pipes

Figure 3.2   Hydraulic jump and plug formations

(a)   conventional;
(b)   bottle;
(c)   integral;
(d)   special.

Conventional traps are formed from a series of bends in the pipe and allow full bore flow through the trap detail. This type of trap does necessitate sufficient space to accommodate the various bends.

In contrast, the bottle trap is a more compact detail which still provides excellent access for the clearance of debris. The working principle of the bottle trap does however cause disruption to the smoothness of the flow through the trap (see figure 3.3) and this prevents its use on the waste leading from sink grinders. Because of the nature of the waste emanating from the sink grinder, accumulation of debris can be expected and this would almost certainly lead to blockage should the bottle trap be used.

Special traps may include those intended to resist siphonage (see figure 3.10) or dilution traps (large body capacity) for use in laboratory work. The dilution trap may be used specifically to dilute a chemical discharge or to facilitate the collection of suspended debris.

Figure 3.3 illustrates examples of the conventional, bottle and integral traps. Figure 3.4 shows an overflow connection to a bath trap which is the detail now generally used for baths.

As retention of liquid in the traps is necessary for the containment of foul air in the pipework, loss of the liquid seal as a result of siphonage effects should be carefully considered. The occurrence of hydraulic jump and plug

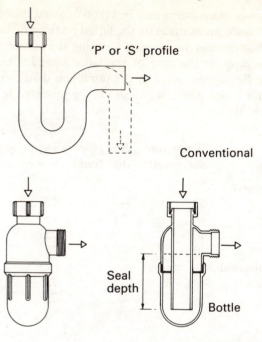

'P' or 'S' profile

Conventional

Seal depth

Bottle

Figure 3.3 Types of trap

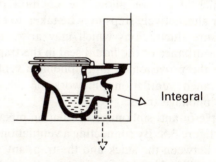

Integral

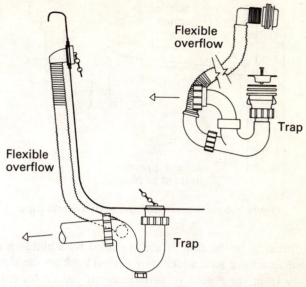

Flexible overflow

Flexible overflow

Trap

Trap

Figure 3.4 Combined trap and overflow to baths

Air compressed in front of jump

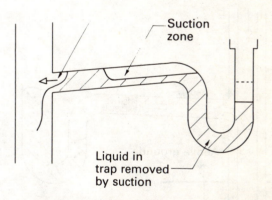

Suction zone

Liquid in trap removed by suction

Figure 3.5 Self-siphonage in a waste pipe

formations may create air pressure fluctuations large enough to break the liquid seal of the trap by siphonage action.

Figures 3.5, 3.6 and 3.7 show how air pressure changes may cause the loss of seal in traps.

Figure 3.5 shows siphonage as a result of the creation of a hydraulic jump formation arising from the use of too small a pipe diameter.

The liquid piston of the jump formation compresses the air in front of it as it moves forward. This causes negative pressure behind the jump and it is this negative pressure which provides the suction force to remove liquid from the trap.

Figure 3.6 shows siphonage of the liquid from a trap which has been induced by the discharges from other appliances further up the stack.

Figure 3.7 shows the effect of back-pressure percolation which is associated with the lower end of stack pipes where entry is made to the below-ground drain. As the contents of the stack gravitate to the invert of the drain pipes (lowest point on the internal bore — see below-ground drainage, chapter 4) there will inevitably be turbulence caused by the change in flow direction. This

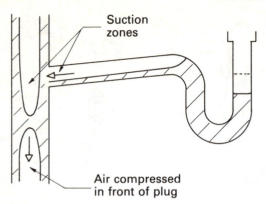

Figure 3.6   Induced siphonage in a waste pipe

disturbance to the flow is accompanied by a build-up of positive air pressure which may extend back up the stack away from the region of turbulence. If discharges from sanitary appliances are connected close to the foot of the stack pipe, the positive pressure may force foul air through the liquid of the trap and into the building.

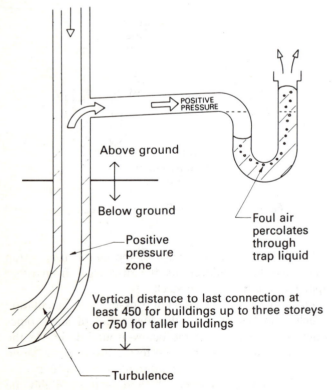

Figure 3.7   Back-pressure percolation

A number of measures may be taken to avoid the effect of air pressure fluctuations on the liquid retained by traps. Self-siphonage may be avoided by using the correct diameter of pipe for the type of appliance served. *Current Building Regulations* refer to the minimum diameters of traps, and hence pipes, which are to be provided to the appliance.[3,4]

Appliance	Waste pipe dia. (mm)	Soil pipe dia. (mm)	Depth of trap seal (mm)
Handbasins	32	—	75
Bidets	32	—	75
Showers	40	—	75
Baths	40	—	75
Sinks	40	—	75
Waste disposal units	40	—	75
Urinals	—	40	75
WC	—	75	50

Induced siphonage may be prevented by using a stack diameter which does not allow the creation of plug formation, or alternatively steps may be taken to equalise any air pressure fluctuations which may arise, thereby preventing disturbance of the liquid seal in the trap. Such steps include the provision of anti-siphonage ventilation pipework, branch ventilation pipework or the use of special anti-siphonage traps.

The principle of anti-siphonage ventilation pipework is illustrated in figure 3.8. By connecting a ventilation source on the waste between the stack and the trap, any suction effects can be nullified by the drawing of air from the vent connection. This of course means additional cost in the provision of the anti-siphonage ventilation pipework system which extends through the building to the external air at roof level.

An alternative to carrying the ventilation pipework throughout the full height of the building to reach the external air at roof level is to loop into the stack pipe in the manner shown in figure 3.9.

The use of this technique may be restricted by other discharges in close proximity which could enter the ventilation loop.

There are forms of bottle trap available which are designed to allow air entry to the waste pipe via the trap under suction conditions but which also maintain a liquid seal once the suction effect ceases. Such a detail is the

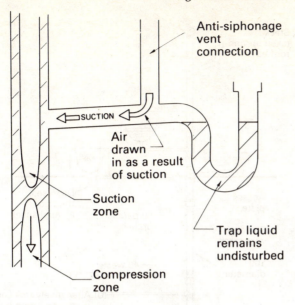

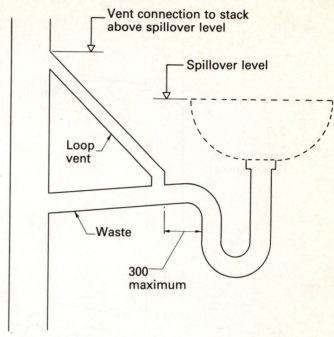

Figure 3.9   *Branch ventilation by loop connection to the stack*

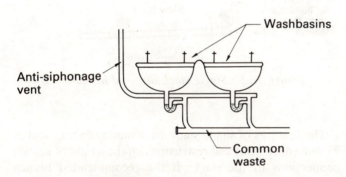

Figure 3.8   *Pressure equalisation by the use of anti-siphonage pipework*

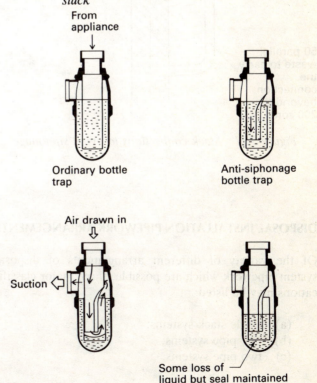

Figure 3.10   *The principle of the anti-siphonage bottle trap*

anti-siphonage bottle trap, the principle of which is illustrated in figure 3.10.

When appliances are connected to a stack there are usually a number of connections to be made on one floor, since bathroom appliances tend to be grouped together. If a waste connection is made to a stack directly opposite a WC connection, the large-capacity discharge from the WC will inevitably cause induced siphonage and loss of seal in the trap served by the waste. Figure 3.11 shows how this effect can be avoided by making connections outside the immediate zone of the WC discharge.

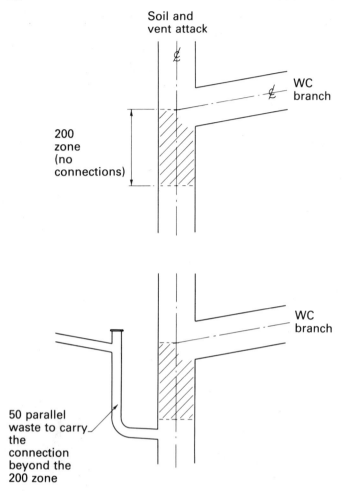

Figure 3.11   Stack connections to avoid siphonage

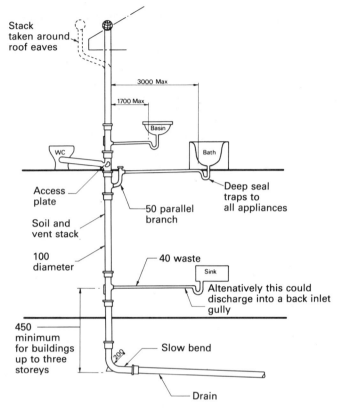

Figure 3.12   Single stack disposal installations

DISPOSAL INSTALLATION PIPEWORK ARRANGEMENTS

Of the variety of different arrangements of disposal system pipework which are possible, three major classifications may be listed:

(a)   single stack systems;
(b)   one pipe systems;
(c)   two pipe systems.

Single stack arrangements are most commonly applied to domestic premises (figure 3.12).

The features of single stack are a minimum trap seal of 75 mm (deep seal) and restrictions on the length of branch connections to the stack. If the recommended branch lengths are to be exceeded, ventilation will have to be provided in the form of anti-siphonage pipework.[4]

A typical one pipe system of collection is illustrated in figure 3.13. As will be seen, this system differs only from the single stack in the provision for full ventilation by anti-siphonage pipework connections.[5]

Two pipe arrangements separate the WC discharges from those of all other sanitary appliances. This results in the use of two vertical stacks — one for soil discharges and one for waste discharges. As shown in figure 3.14, the system may also be provided with anti-siphonage pipework connections.

When choosing one of the disposal pipework arrangements, the proposed layout of the appliances in the building to be served would be the major influential factor.

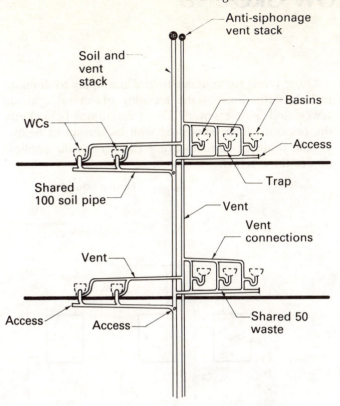

Figure 3.13   One pipe disposal installations

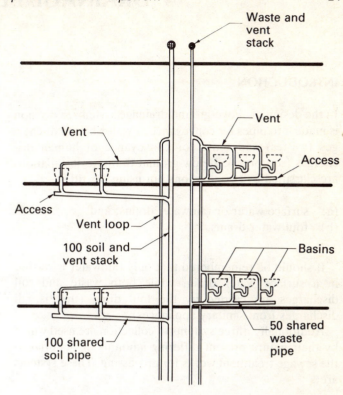

Figure 3.14   Two pipe disposal installations

Provided careful design is employed, the single stack system, which is an economic solution in terms of pipe length, may be used on multi-storey construction as well as low rise structures. This does however mean that the appliances will have to be grouped around the stack, and in situations where such grouping is not possible the one pipe or two pipe system may be most appropriate.

REFERENCES

1 See *BS 5572: 1978 Code of practice for sanitary pipework (formerly CP 304); Building Regulations 1985,* Part H, clause H1 (1), Approved Document H, Section 1, Sanitary pipework.
2 *Building Regulations 1985*, Approved Document H, Table 1 to H1, Flow rates for sanitary appliances.
3 *Building Regulations 1985*, Approved Document H, Section 1, Traps.
4 *Building Regulations 1985*, Approved Document H, Section 1, Branch discharge pipes.
5 See *Building Regulations 1985*, Approved Document H, Appendix to H1, clause A6.

4 DRAINAGE BELOW GROUND

INTRODUCTION

In the design of above-ground drainage systems, a division is made into pipes for conveyance of water-bound discharges (wastes) and those for conveyance of human discharges (soil pipes). Below-ground drainage installations are classified by two divisions but using the titles:

(a) surface water or rainwater drains; and
(b) foul water drains.

It should be remembered that only rainwater is carried in a surface water drain, while both waste and soil discharges are carried by the foul drain (that is, all discharges from sanitary appliances).

Additionally, three systems of collection are used which by their nature present differing amounts of material to the sewage treatment works for processing. These systems are:

(i)   the combined system;
(ii) the separate system; and
(iii) the partially separate system.

In the combined system, rainwater and foul water discharges are mixed together and this means that an unnecessary burden is placed on the sewage treatment works. Recognising the benefits of separating rainwater from other discharges, the application of a separate method of collection is generally preferred.

A variation on the combined and separate systems is the partially separate system. In this the rainwater collected from public roads and other public paved areas is carried by the surface water drain, while the rainwater from private roofs and paved areas is discharged into the foul drain. A division of classification into public drainage and private drainage can be made in this instance.

Other terms to be understood which apply to drainage include definitions of the meaning of 'drain', 'private sewer' and 'public sewer'. These definitions help clarify the responsibilities associated with ownership of below-ground drainage runs. Figure 4.1 illustrates the application of these terms.

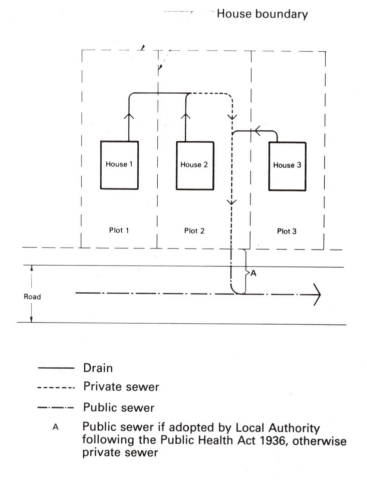

Figure 4.1   Drains, private sewers and public sewers

From a design point of view, a number of features may be incorporated which will ensure efficiency in operation:

(1) Pipe layouts should be simple, with as few changes in direction or gradient as possible.
(2) Pipe runs should be in straight lines.
(3) The system should be ventilated (one of the functions provided by the above-ground drainage stack pipe).
(4) Pipes should be laid at a sufficient depth to avoid damage.
(5) Pipe gradients should suit the material being conveyed — flows containing solids and liquids should achieve a self-cleansing velocity.
(6) Sufficient capacity should be provided by the use of pipes of diameter to suit the peak flow.
(7) Good quality pipe materials should be used (to the relevant British Standard), and the material used should suit the ground conditions.
(8) The pipe joints must be watertight — flexible joints generally allow a 5° distortion of one pipe relative to another without breaking the joint seal.
(9) Suitable bedding and backfilling should be used to the pipes.
(10) Sufficient points of access should be provided for testing and the clearing of blockages.
(11) Inlets to the drain should be trapped (except for the entry at the foot of the soil and vent stack).
(12) Junctions to the main flow should be minimised and should be made in the direction of the main flow.[1]

Adherence to this list should provide a system which will function with little need for maintenance.

In conveying its contents to the point of termination, which may be the Local Authority sewer, the drain should be inclined to achieve a self-cleansing velocity. This in turn will dictate the depths of pipe trenches and the overall economy of the excavation.

When sizing the pipe diameters for a system, an estimate may be made of the extent of the flows emanating from the sanitary appliances, but the extent of rainwater to be accommodated may be less predictable. If we divide the below-ground drains into surface water drains and foul water drains, the features considered in sizing for overall capacity can be examined. Individual surface water drains will be sized in relation to:

(a) the catchment area of the ground that they take a discharge from;
(b) the intensity of rainfall over this area (perhaps assessed by reference to meteorological records);
(c) consideration of the permeability or impermeability of the ground within the catchment area;
(d) the gradient of the pipe diameter proposed which will determine the flow velocity and the rate of flow.

The factors considered when sizing foul water drains may include:

(a) the quantity of discharge from the sanitary appliances served by the drain (their capacity and frequency of use);
(b) allowance for periods of peak usage;
(c) the gradient of the pipe diameter proposed which will determine the flow velocity and the rate of flow.

BS 8301: 1985 Code of Practice for Building Drainage contains a number of tables to assist in the correct sizing of drain pipes. Of particular use in this task are:

(i) Table 4 — Flow rates, probability of discharge factors and discharge unit ratings.
(ii) Figure 1 — Probability graph for number of appliances discharging simultaneously.
(iii) Figure 2 — Design flows for foul drains: conversion of discharge units to flow rates.

These desirable features are now discussed under the headings of inlets to the drain, access, pipe materials, jointing, bedding and termination of the drain.

INLETS TO THE DRAINS — GULLIES AND TRAPS

Most drain runs will commence immediately outside the perimeter walls of the building and to prevent foul air from the drain escaping into the air around the building, the gully detail is employed.

The seal which is obtained by the use of a gully is one which is formed by the use of a bend in the same manner as the seal provided by traps to sanitary appliances.

Traditionally the gully is a one piece detail but, as shown in figure 4.2, the two piece gully has the advantage of manoeuvrability in that its trap may be rotated to suit the direction of the drain run which is to follow.

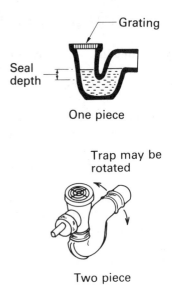

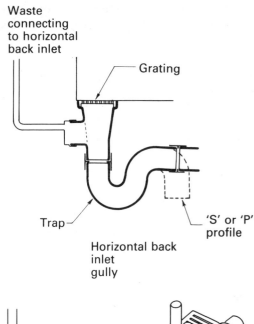

Figure 4.2   One and two piece gullies (two piece gully) courtesy of Hepworth Iron Co. Ltd)

Such gullies may receive the discharges from rainwater pipes or the waste pipes from sanitary appliances. In the latter case *Current Building Regulations* stipulate that the waste pipe must discharge its contents below the level of the grating but above the trap water line.[2] This may be achieved by the use of a back inlet gully — either with a horizontal or vertical back inlet (figure 4.3).

The same basic profile of the gully is reproduced in the larger yard gully detail, and in the road gully which is illustrated in figure 4.4.

The road gully illustrated in figure 4.4 is the profile section provided for clayware details. Gullies may also be obtained in concrete or cast iron.

The depth and capacity of road gully details prevent the use of lift-out perforated buckets for sediment collection and necessitate the use of Local Authority suction tankers for this purpose.

Special details may also be used which have an intercepting role, for example in the interception of sediment or petrol. Figure 4.5 shows a mud gully detail which has a lift-out sediment pan.

The petrol interceptor is a detail generally employed on the forecourt of petrol filling stations and the like. This works on the principle that petrol, being less dense than water, will float on the surface of water and evaporate. By carrying the petrol-contaminated discharge through a

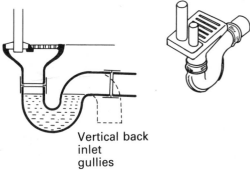

Figure 4.3   Horizontal and vertical back inlet gullies (courtesy of Hepworth Iron Co. Ltd)

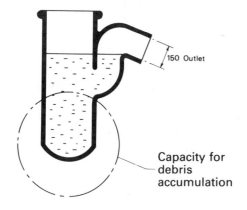

Figure 4.4   A typical road gully

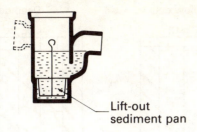

Figure 4.5   A mud-intercepting gully

series of chambers which have an overspill well below the chamber water line, a less contaminated flow should progress to the next chamber. A vent pipe is used to collect the petrol vapour and to transfer it to a point where it can be safely discharged into the open air, as shown in figure 4.6.

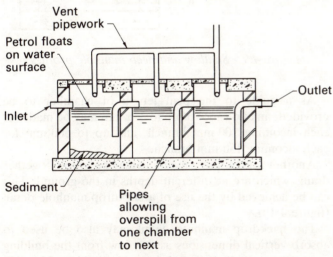

Figure 4.6   The petrol interceptor

As an alternative to the *in-situ* brickwork detail shown, the petrol interceptor may also be obtained in pre-formed glass reinforced plastic (GRP — fibreglass).

ACCESS TO THE DRAIN

The details used to provide access to a drain may be listed as:

(a)   manholes;
(b)   inspection chambers;
(c)   rodding eyes;
(d)   access fittings;
(e)   roddable gullies.

Traditionally constructed manholes consist of brick-work and concrete as shown in figure 4.7. This detail shows the common practice of exposing the flow as it enters the chamber by the use of a half-section pipe referred to as the channel. Any spillage is redirected towards the channel by the surrounding concrete bench-ing. This detail clearly shows the term 'invert level', a term which is of fundamental importance in the design of the system in ensuring the provision of the correct falls to the pipes.

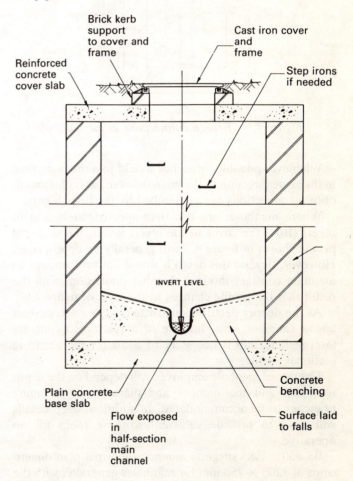

Figure 4.7   Traditional brick manholes[3,4]

Manholes are not only used to inspect the flow but are positioned at points where flows are brought together since such points are the site of potential blockages.[5]

To aid the turning of branch connections to the main drain run, special threequarter-section branch bends are used (figure 4.8).

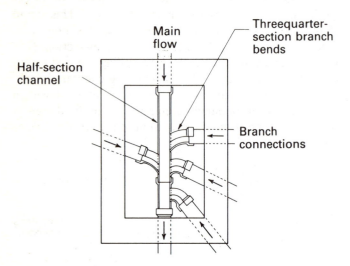

Figure 4.8   Branch connections at the manhole

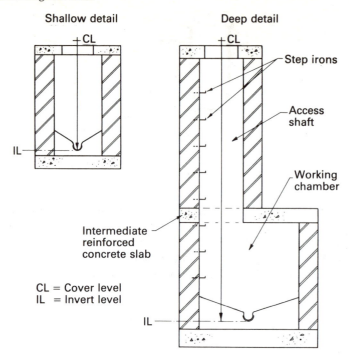

Figure 4.9   Shallow and deep manhole profiles

Whenever possible, branches should join the main flow in the same direction as the main flow but, as is illustrated, obtuse connections can be handled by the branch bend.

Where manholes are less than approximately 2 m to invert (from the cover to the invert level), the sectional profile shown in figure 4.7 will generally be appropriate. However, beyond this depth it would not be economic to maintain constant internal chamber dimensions with the result that the profile changes to the style of figure 4.9.

As the deeper detail is constructed with an access shaft and a chamber area, the use of a reinforced concrete intermediate slab is necessary to allow for the changing sectional profile.

The size of manhole employed will depend on the depth to invert and the number and diameter of incoming branches to be accommodated. In addition, deep details will have to provide sufficient working room for an operative.

BS 8301: 1985 suggests minimum internal plan dimensions of 1200 × 750 mm for manholes generally, with the deeper chambers served by an access shaft of minimum internal dimensions 900 × 840 mm (see figure 4.9).

As a guide to the plan length of chamber to be provided, an allowance of up to 375 mm may be made for each incoming 100 mm branch, and up to 500 mm for each incoming 150 mm branch.

Another function of a manhole may be to join together drains which are at different depths in the ground. This can be achieved by the use of a backdrop manhole detail (figure 4.10).

The backdrop manhole detail may also be used to absorb vertical dimensions as the flow from the building moves towards the point of drain termination. This will allow desirable pipe gradients to occur between the back-drop chambers and preservation of a self-cleansing velocity for the flow.

To achieve vertical movements of the flow using a backdrop chamber in this way will be particularly appropriate where the inclination of the ground results in a considerable difference in the levels of invert between the points of origin of the drain at the building, and the point of termination at the Local Authority sewer.

BS 8301: 1985 suggests that in situations where the difference in invert levels of the flows being joined together is less than 1 metre, that a ramp can be formed

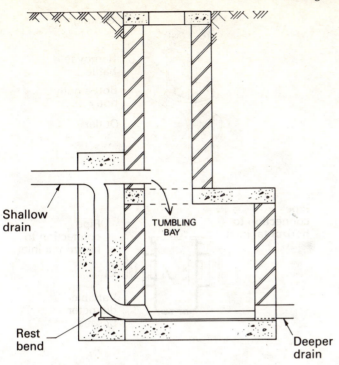

Figure 4.10   The backdrop manhole

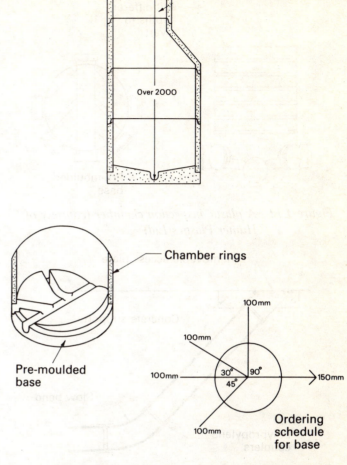

Figure 4.11   A pre-cast concrete manhole

between the upper drain and lower drain at the manhole position with pipes inclined at around 45°.

The tumbling bay shown allows overspill to occur in surcharge conditions and also provides a means for rodding the incoming drain.

Manholes are also available in pre-cast concrete. As illustrated in figure 4.11, a pre-moulded base is commonly supplied with the desired number of connections. This figure also shows a typical ordering diagram.

The pre-moulded base is cast with its upper surface to falls, and this eliminates the need for *in-situ* concrete benching around the channel and branches. When laying pre-cast concrete manholes, it is usual to encase the assembled chamber in a thin jacket of *in-situ* concrete for stability.

Current Building Regulations require manhole details to occur at least every 90 m or every 45 m where measuring between a manhole and an inspection chamber.[5]

Inspection chambers may be constructed of the same materials as a manhole but are physically smaller units.[6] By definition, a manhole has a chamber of volume capable of accommodating a man whereas it is the inten-

tion that inspection chambers should allow inspection or rodding of the flow from ground level. The plastic inspection chamber is a detail commonly applied (figure 4.12).[7]

Rodding eyes are normally used as an alternative to the use of a manhole or inspection chamber for clearing blockages where two flows come together. This access facility consists of a removable plate cover attached to the end of a pipe which is brought up to ground level by the use of a slow bend (figure 4.13).

The one and two piece gullies mentioned earlier have traditionally been disregarded as a means of access to the drain for the rodding of blockages since the bend of the gully is too sharp to allow entry of the rods. A plastic bottle gully is however now available (Wavin Plastics — Osmadrain) which has a removable interior

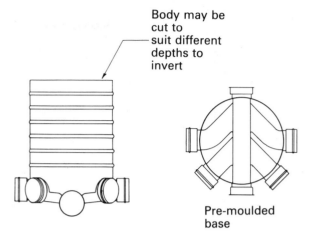

Figure 4.12   *A plastic inspection chamber (courtesy of Hunter Plastics Ltd)*

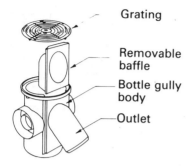

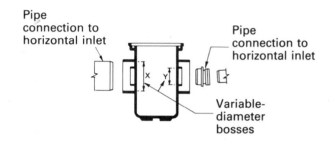

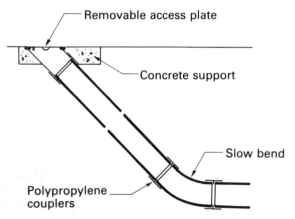

Figure 4.13   *A rodding eye detail (courtesy of Hepworth Iron Co. Ltd)*

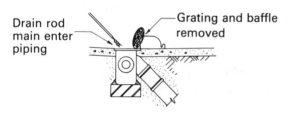

Figure 4.14   *A roddable gully (courtesy of Wavin Plastics Ltd — Osmadrain)*

enabling the following drain to be rodded (figure 4.14). This gully may also be used with horizontal inlets as a back inlet detail.

PIPE MATERIALS, JOINTING AND BEDDING

Clay pipes have been used in below-ground drainage for many decades. In their original form the clay was rendered impervious by the technique of salt glazing and the pipes were referred to as 'salt glazed ware'. Recent developments in the manufacture of drainage goods have revolutionised the process by which the naturally porous material is made impervious and the technique now applied is vitrification.

Although vitrified clay appears visually porous because of its dull finish, the internal characteristics of the clay have been changed and become glass-like under the heat treatment process. Specification of clay pipes is now by use of the abbreviation GVC or glazed vitrified clayware.

Clay pipes and fittings are currently produced by BS 65[8] quality standards which allows jointing in a number of ways.

Socket and spigot joints are traditional but plain ended pipes can be used which have special polypropylene flexible joints (figure 4.15).

Other common pipe materials include:[9]

concrete   BS 5911
asbestos   BS 3656

iron   BS 437
pitch fibre   BS 2760
uPVC   BS 4460 and BS 5481

When pipe materials are considered, they are usually grouped under the headings of rigid pipes and flexible pipes. The choice of material will be made on the need to

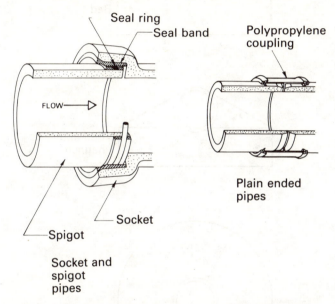

Figure 4.15 Socket and spigot, and plain ended pipe joints

provide adequate strength to resist loading deformation or damage, the nature of the discharge conveyed, and the prevailing ground conditions. Clay pipes to BS 65, like other rigid pipes, do not deform noticeably under load. Three strengths are listed in the British Standard, and pipe lengths supplied are up to 3 metres. Like so many natural building materials, clay is resistant to most substances which could be found in the flow or surrounding soils. There are, however, special chemically resistant pipes also available.

BS 5911 covers both plain and reinforced concrete pipes. As with any material which is cement based, care has to be exercised in preventing damage under the action of sulphates. Supplied lengths are up to 2.50 metres.

Asbestos–cement pipes are particularly brittle and rely heavily on the provision of correct bedding to prevent fracture. Again, this type of pipe should not be used in

sulphate-containing soils. Supplied lengths are commonly 3, 4 and 5 metres, with $\frac{1}{2}$ and $\frac{1}{4}$ pipes available to aid jointing.

Iron pipes to BS 437 are termed 'grey iron' and classed as rigid pipes because of their brittle properties. Here protection from corrosion is important, and with this in mind acidic soils such as peat should be avoided. Grey iron pipes are for use with either caulked lead or flexible joint couplings. Supplied lengths are up to 5.5 metres.

Flexible pipe materials include pitch fibre, uPVC, GRP and steel. Pitch fibre has the advantages of flexibility and as a consequence may be fairly easily manoeuvred. They should not however be used to convey flows which are hotter than 60°C. Supplied lengths are up to 3 metres.

uPVC pipes similarly are only suited to flows of up to 60°C. This pipe material has the advantage of very low dead weight in respect to handling, but has the disadvantage that it tends to become brittle at low temperatures. Supplied lengths are up to 9 metres.

Glass reinforced plastic (GRP) pipes are equally affected by high temperature flows and the advice of the manufacturer should be sought where hot flows are anticipated. Being a very light material, the lengths supplied are up to 12 metres.

Steel pipes fall into the flexible classification since they do allow some flexibility. Special attention may be needed here though, in relation to corrosive elements which may be contained in the flow or the surrounding soil.

Table 3 of BS 8301: 1985 should be consulted for more specific information on the performance of different pipe materials in terms of chemical resistance. Irrespective of the pipe material chosen, there are two basic joints employed, namely rigid joints and flexible joints. Some rigid and flexible joints are illustrated in figure 4.16.

A general feature of the flexible joint is its ability to allow a 5° distortion between one pipe and another while maintaining a seal. Other advantages may also be listed as follows:

1. flexible joints are dry joints and this means that the drain can be immediately tested;
2. less skill is required in forming the joint and jointing can consequently be done more quickly;
3. poor weather does not interrupt laying.

The bedding given to drainage pipes may be of

(a) concrete;

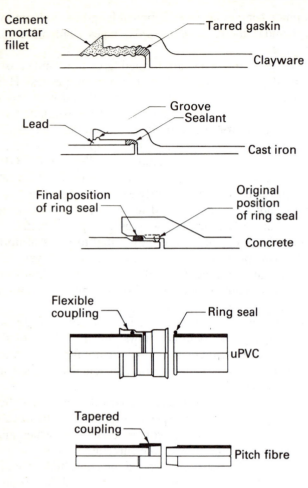

Figure 4.16   A selection of rigid and flexible joints

is limited to high load situations. Traditional concrete beds, beds and haunchings, and beds and surrounds are illustrated in figure 4.17.

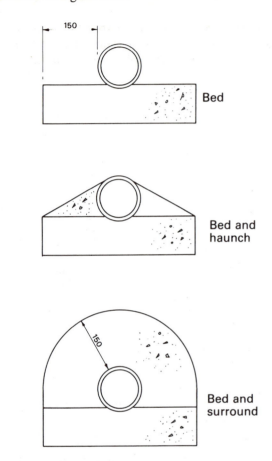

Figure 4.17   Traditional concrete supports to pipes

(b)   granular material;
(c)   concrete with provision for movement;

or alternatively

(d)   the pipes may be bedded directly on the earth of the trench bottom.

The nature of the support given to the pipes will considerably influence the overall strength of the drain. Bedding pipes directly to earth requires very high standards of workmanship since any voids occurring below the pipe body will detract significantly from its strength.

Concrete is the traditional bedding medium for drain pipes but its applicability in the *Building Regulations 1985*

Granular beds may be provided to rigid or flexible pipes, and in the case of the latter a backfilling of granular material to the top of the pipe is normally recommended. Granular aggregates are usually BS 882 quality.

In ground where movement or heavy loading is anticipated, rigid pipes should be encased in concrete. The integrity of flexible joints may be preserved by the building in of compressible material at the pipe joint as shown in figure 4.18.

When flexible pipes are used in ground where movement is anticipated a granular bedding and backfill is generally prescribed, with the backfill extending to at least

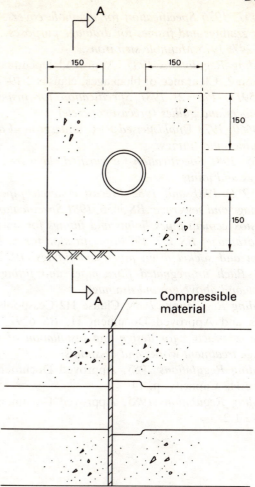

Figure 4.18   Movement joints in concrete pipe encasement

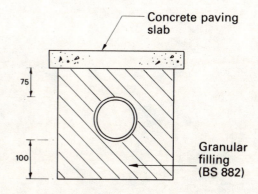

Figure 4.19   Granular encasement for flexible pipes

75 mm above the pipe. The covering of the encased pipe with a paving slab will provide additional protection from compression loads from above (figure 4.19).

DRAIN TERMINATION

The introduction to this chapter considered the populated environment where drains discharged into sewers for conveyance of effluent to the sewage treatment works. Less well populated areas however may not be served by Local Authority sewers and a cesspool[10] or septic tank detail may be employed. Cesspools are simply holding chambers which will be periodically emptied by the Local Authority.

Septic tanks provide a crude form of sewage treatment before the contents are discharged into a nearby water course.

CESSPOOLS

As it is estimated that a three-person household will generate approximately 7 m³ of discharge in about three weeks, and the *Building Regulations*[11] recommend a minimum chamber capacity of 18 m³, many cesspool details will require the Local Authority to make frequent visits for the purpose of sewage collection. BS 6297 suggests that the design should accommodate at least 45 days' discharge.

In addition to capacity, other constructional features will include shape (cylindrical chambers are often recommended), and the need to ventilate. Foul air ventilation will usually be provided by the soil and vent stack of the building drainage system, while provision should also be made for a fresh air inlet fitted with a non-return mechanism.

SEPTIC TANKS

The minimum capacity suggested for septic tanks is 2720 litres, and a formula for sizing the total capacity of the detail is given in BS 6297:

$$C = (180P + 2000)$$

where C is the capacity in litres and P represents the population being served.

As sewage passes through a septic tank a delay is caused by movement of the material through successive chambers (normally two). During this delay, partial purification of the sewage occurs as a result of the generation of anaerobic bacteria. In this natural process, the by-products are methane gas, sludge and liquid. As with the cesspool detail, it is important to ventilate, and here a safe position needs to be found for the ventilation of the flammable methane gas. Since much of the sludge will accumulate as the effluent enters the first chamber, this will normally be sized to accommodate two-thirds of the total capacity of the detail.

Both cesspools and septic tanks must be constructed to ensure that seepage does not occur either from or into the chambers. Suitable body materials include high density brickwork (engineering quality), concrete, steel and glass reinforced plastic (GRP).

REFERENCES

1 *Building Regulations 1985*, Approved Document H, Clause H1, Section 2, Pipe gradients and sizes, clauses 2.10–2.14, Bedding and backfilling, clauses 2.16–2.18.

2 *Building Regulations 1985*, Approved Document H, Hl Section 1, Branch discharge pipes, clause 1.11.

3 *BS 497: 1976 Specification for manhole covers, road gully gratings and frames for drainage purposes.*

4 *BS 1247: 1975 Manhole step irons.*

5 *Building Regulations 1985*, Approved Document H, Section 2, Clearance of blockages, clauses 2.19–2.26.

6 *BS 5911: Part 2: 1981 Specification for inspection chambers and gullies (precast concrete).*

7 *BS 4660: 1973 Unplasticised PVC underground drainage pipes and fittings.*

8 *BS 65: 1981 Specification for vitrified clayware pipes, fittings and joints.*

9 *BS 5911: 1981 and 1982 Precast concrete pipes for drainage and sewerage; BS 3656: 1981 Specification for asbestos–cement pipes, joints and fittings for sewerage and drainage; BS 437: 1978 Specification for cast iron spigot and socket drain pipes and fittings: BS 2760: 1973 Pitch impregnated fibre pipes and fittings for below and above ground drainage.*

10 *Building Regulations 1985*, Clause H2 Cesspools and tanks and Approved Document H; *BS 6297: 1983 Code of practice for design and installation of small sewage treatment works and cesspools.*

11 *Building Regulations 1985*, Approved Document H, clause H2 Capacity, p. 14.

12 *Building Regulations 1985*, Approved Document H, clause 1.2.

5 SPACE HEATING OF BUILDINGS

INTRODUCTION

There are numerous forms of heating system which may be used for space heating. A factor very influential in the choice today is costs: costs not only in terms of the initial installation but also in terms of fuel consumption and maintenance. Before selecting a system, examination of both initial and running costs needs to take place to provide a total cost picture over the projected life of the building.

Traditionally water has been the medium through which heat has been conveyed from the boiler to the point of emission, and this continues to be the case for residential heating. This chapter examines the methods of heating for both residential and non-residential property.

LOW-PRESSURE HOT WATER HEATING[1]

The title relates to pumped hot water systems which may use a variety of fuels in hot water generation — solid fuels, oil and gas. The pipe layouts to the emitting radiators are basically of two forms:

(a) one pipe arrangements; or
(b) two pipe arrangements.

Typical layouts in one pipe systems include the 'ring', 'ladder', 'drop' and 'parallel' systems (figures 5.1 and 5.2).

The water in the one pipe system is conveyed around the heating circuit effectively in series with flow temperatures typically 80°C with an overall temperature drop in the region of 10–20°C on return to the boiler. Because the system is effectively in series, the temperature of the water will fall progressively as it enters successive radiators and this may cause some difficulty in balancing the heat output in different areas.

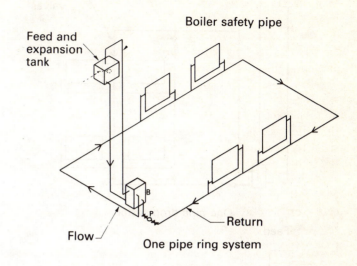

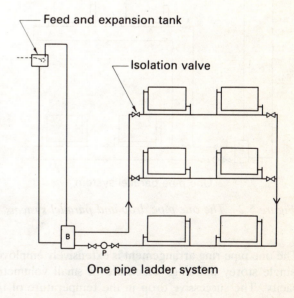

Figure 5.1 *The one pipe ring and ladder systems*

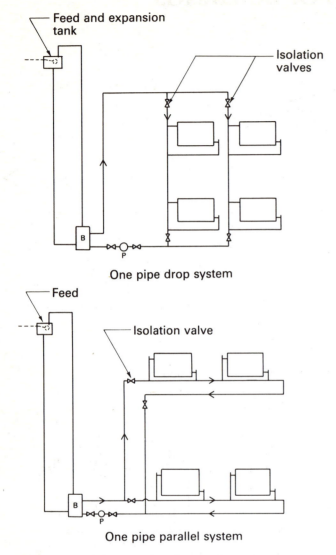

Figure 5.2   The one pipe drop and parallel systems

The one pipe ring arrangement is extensively employed in single storey buildings which are of small volumetric capacity. The successive drop in the temperature of the flow will dictate restriction to buildings of small volume.

In contrast, the one pipe ladder, drop and parallel systems are all suitable basic designs for buildings of more than one storey. Ladder systems involve the main distribution of heated water laterally to feed banks of radiators on each floor level. Here exposure of the horizontal feed pipe may also add to the radiant heat provided to the room space. With the drop system the hot water feed to

the radiators is supplied from a horizontal distribution pipe located at the top of the building. Vertical feeds from this will serve one radiator on each floor level. The parallel system provides a single pipe for the flow and return to all the radiators on each of the floor levels.

HEATING SYSTEM CAPACITY

When designing the heating system for a building the fundamental considerations are the internal room temperatures to be maintained, the rate of heat loss from the rooms and the extent to which heat may arise from other sources (either as a result of occupancy or solar gains).

The temperature required in the various room areas can be set and the rate of heat loss determined. In this way the output of the radiators can be calculated to suit these two parameters. The temperature of the flow in low-pressure hot water heating systems is, as already mentioned, generally around 80°C with a 10–20°C drop anticipated. By assessing the temperatures needed at the heat emitter in each room and the drop in flow temperature, the size of the distribution pipework can also be assessed.

Other influential factors in the sizing of pipes in low-pressure hot water heating systems are the water flow rate required and the pressure losses experienced by the flow as a result of friction. Tables are available to assist in this exercise, which relate pipe diameter to both flow rate and pressure loss.

A preferable solution to a one pipe system from a heat balancing point of view is likely to be a two pipe layout. Typical layouts in two pipe systems include the 'drop', 'parallel', 'upfeed', and 'high-level return' systems (figures 5.3 and 5.4).

With two pipe systems a separate flow pipe and return pipe is used which means that effectively the water is carried around the system in parallel, making the balancing of radiator outputs less difficult. With the two pipe drop arrangement a major advantage is that air does not tend to accumulate in the radiators since the layout encourages air in the system to rise to the high-level distribution pipe from where it can be vented.

Two pipe parallel systems provide a flow and separate return along each floor level in the building. By the provision of a separate flow and return, the balancing of radiator temperatures can be more easily achieved, this point applying to all two pipe systems.

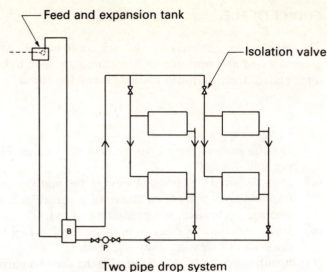

Two pipe drop system

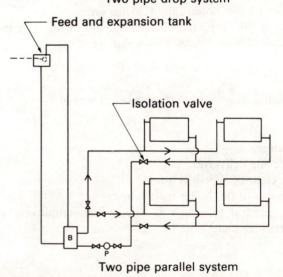

Two pipe parallel system

Figure 5.3 *The two pipe drop and parallel systems*

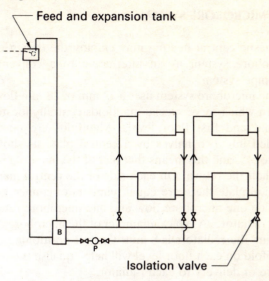

Two pipe upfeed system

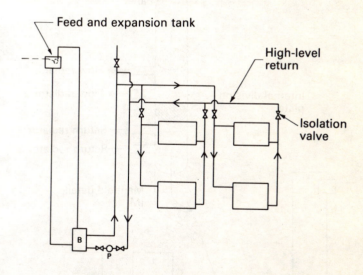

Two pipe high-level return system

Figure 5.4 *The two pipe upfeed and high-level return systems*

When different areas within the building contain differing numbers of storeys, the use of an upfeed arrangement as shown in figure 5.4 provides for all the flows to emanate from floor level. This means that vertical rises can be easily varied to suit the number of floors to be served.

In contrast, to have both the flow and return at high level, as in the high-level return system, frees the lower floor of the need to accommodate a main horizontal pipe. This may be particularly useful where the use of a low-level return would necessitate building into a solid floor.

THE MICROBORE SYSTEM

Domestic central heating may employ the design of the microbore system as an alternative to a conventional one pipe system.

The microbore system uses a 22 mm or 28 mm flow and return which serve special manifolds centrally located on ground and first floor levels. Manifolds are internally divided into two halves by a central plate as shown in figure 5.5, and this means that half of the manifold is flow and half the return. On each side of the central manifold division plate there are equal numbers of connections to provide one microbore flow and one microbore return to each radiator. A typical diameter of the microbore circuit feeding the radiators is 8 mm. Central positioning of the manifold on each floor level will help equalise the temperature of delivery to each radiator.

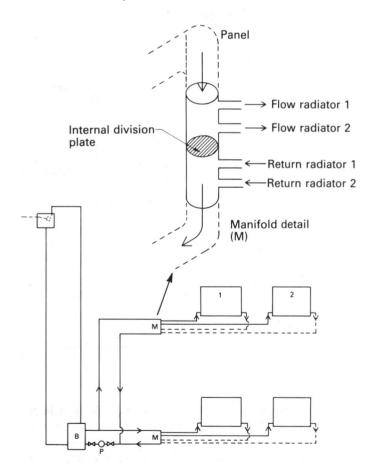

Figure 5.5   The principle of the microbore system

CHOICE OF FUEL

The main choice criteria in the selection of fuel are generally cost and convenience, and these are seen to be interrelated. Cost itself will embody many factors:

(a)   current supply costs;
(b)   the efficiency of the fuel in terms of the heat energy per unit of cost;
(c)   from an initial cost point of view — the implications on building design of the choice of a particular fuel (storage enclosures, fire precautions, etc.);
(d)   labour attendance costs — perhaps feeding fuel to the boiler or clearing away ash residue;
(e)   maintenance costs associated with the use of a particular fuel — boiler maintenance etc.;
(f)   the costs incurred in complying with legislation — Factory Acts, Clean Air Acts, etc.

Convenience may involve the consideration of:

(i)   storage;
(ii)   ordering/deliveries;
(iii)   the clearing of ash residue;
(iv)   the size of the boiler installation;
(v)   the general reliability of the system employing a particular fuel.

Fuel Burning

It should be recognised that the efficiency of burning will not only relate to the type of fuel but to the manner in which it is presented to the point of combustion.

Bituminous fuels such as coal or coke are fed to the point of combustion in 'lump' or powder form. Lumps may have to be graded for some feed systems and powdered fuels are usually associated with the higher rated boiler systems.

For feeding the fuels a number of automatic stoking devices are available. Figure 5.6 illustrates the principle of the coking stoker, the underfeed stoker,[2] and the chain grate stoker.

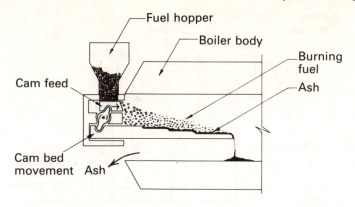

The coking stoker

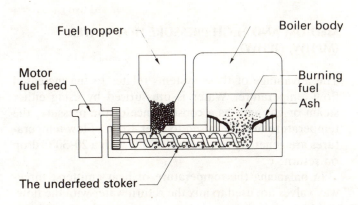

The underfeed stoker

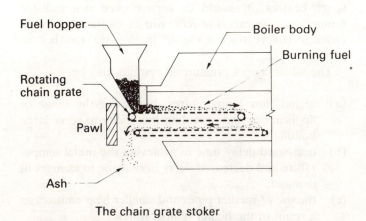

The chain grate stoker

Figure 5.6 Automatic stoking devices

When oil is burnt for water heating it is delivered to the point of combustion in atomised form to aid mixing with oxygen, thereby ensuring good combustion. Monitoring levels of carbon dioxide (CO_2) is the means by which good combustion is assured. Jets to achieve the atomised delivery of oil are varied in design; figure 5.7 shows two types of jet.

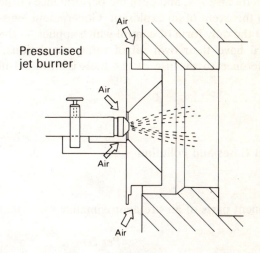

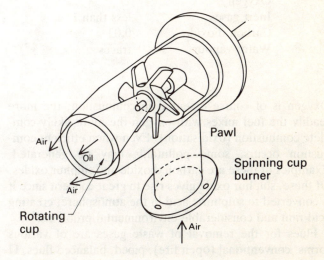

Figure 5.7 Oil atomisation delivery jets

One consideration in relation to oil storage is allowance for fracture of the storage vessel. The construction of concrete or brickwork bund wall enclosures are specifically for this purpose. Such enclosures have volumetric capacities equivalent to the capacity of the tank that they serve.

In relation to the use of gas fuels, safety is the predominant factor. There is a considerable quantity of legislation in respect to safety which applies to supply pipework, distribution within the building, the safety features of the burner units themselves, and even the performance of the structure in the event of an explosion. Government legislation is not the only form of control which applies — the Gas Council have their own set of safety regulations.[3] Further reference to the use of gas fuels is included in chapter 8.

Combustion Gases and Pollution

The component parts of air are approximately

	Per cent
Nitrogen	78
Oxygen	21
Inert gases	less than 1
Carbon dioxide	0.03
Water vapour	traces

Oxygen is of course essential in combustion; the more readily the fuel mixes with oxygen the more likely complete combustion (oxidisation). Even when efficient combustion occurs, some pollutants may be generated. Examples of these are carbon dioxide and sulphur oxides. Of these, sulphur oxide gives rise to great concern since it is converted to sulphuric acid in the atmosphere, creating acid rain and considerable environmental problems.[4]

Flues for the removal of waste gases are of various forms: conventional (open fire), piped, balanced flues, U ducts, SE ducts, etc. These are discussed in chapter 8. Expulsion of gases may be by natural convection (warm air rising) or by fan-assisted arrangements as shown in figure 5.8.

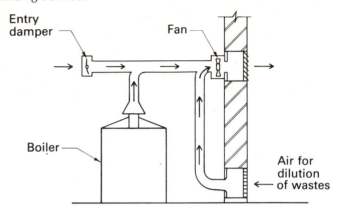

Figure 5.8   Fan-assisted removal of combustion gases

MEDIUM AND HIGH PRESSURE HOT WATER (MPHW, HPHW)

The advantage of these systems relates to the increased flow temperature. Water is pressurised by using either steam or gas and, as a consequence of the pressure, the temperature may be raised to over 100°C. Flow temperatures are generally in excess of 120°C with a 20–30°C drop on return.

In balancing the temperature of heat emitters, three-way valves are used to mix the return water with the flow.

Although high temperatures are obviously desirable for space heating, it should be remembered that radiator temperatures in excess of 80°C will necessitate the use of radiator covers since radiator bodies are too hot to expose.

The advantage of pressurised systems may be:

(a) higher flow temperatures, increasing the range of applications and providing the ability to serve large buildings;

(b) decreased delay time in achieving the initial temperature and decreased delay in reacting to changes in demand;

(c) the use of smaller pipes and smaller heat emitters as a result of the higher temperatures.

Disadvantages may include:

(i) the need for 'soft' water to reduce scaling on the boilers and equipment;

(ii) the need for high-quality materials and workmanship because of pressurisation;

(iii) the need for specific precautions to be taken in relation to pipe expansion and pressure relief.

When comparing low, medium and high pressure hot water heating systems, the following may be listed:

System	Flow temp. approx. (°C)	System pressure
Low pressure	80	Gravity or low-pressure pump
Medium pressure	120+	300 kPa+
High pressure	150+	600 kPA+

(1 Pascal = 1 N/m^2)

STEAM HEATING

The resistance to the adoption of steam heating in the UK has largely been a result of the economics of this system. Steam tends to be associated with high initial costs, high maintenance costs and high insurance costs.

Of the variety of systems available, three forms emerge: gravity, mechanical and vacuum. Systems may also be classified by pressure: low pressure, medium pressure and high pressure. The means of collection of condensate is a way to distinguish between gravity and mechanical

systems. In the gravity arrangement, the condensate will run back to the boiler under natural gravitational force whereas, in the mechanical arrangement, condensate is collected and pumped back to the boiler. Vacuum systems are used to produce steam by boiling water at less than 100°C in reduced pressure conditions.

The typical component parts of a simple system are illustrated in figure 5.9. No pumps are necessary with these systems because of the natural velocity which steam possesses.

ELECTRIC HEATING

The lack of popularity of electric heating for space heating arises from its relatively expensive running costs. This factor seems to outweigh possible advantages which could be listed for this type of heating as:

(i) quick response on activation;
(ii) no combustion wastes to expel;
(iii) ease of temperature control;
(iv) low maintenance needs.

As a generalisation, electric heating may be divided into 'direct' and 'storage' systems. The simplest example of the direct system is a bar radiant wall heater, and of the storage system the night storage heating installations common in residential property.

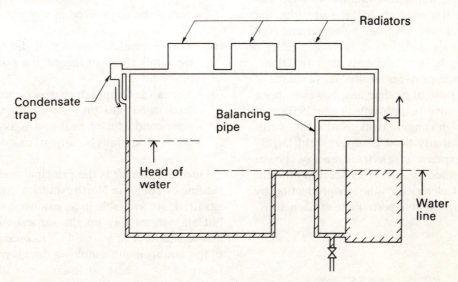

Figure 5.9   The basic components of a steam heating system

Thermal storage heating is an attempt to make electric space heating more competitive by the use of off-peak electricity for which special low rate 'white meter' tariffs are charged. Electricity is consumed on white meter installations between 7 pm and 7 am. The heat generated is stored in high-capacity thermal blocks until it is released by the consumer. Despite the advantages of cheaper off-peak tariffs, storage heating may be still considered as having some disadvantages because of:

(a) Inflexibility — once the electricity has been consumed during the off-peak hours, the cost of the fuel has been incurred. Should a change in the weather eliminate the need for space heating on a particular day, the heat is already in the blocks and no financial benefits can be gained by switching off the system.
(b) The storage heater units are often bulky and, although movable, are extremely heavy.
(c) Economic use of electric space heating is currently only associated with very high levels of thermal insulation.

Block storage heaters contain electric heating elements embedded in high capacity concrete or ceramic blocks inside insulated steel casings. The consumption capacity is generally 1.5–3.0 kW, and this will heat the contents to temperatures in excess of 150°C.

Certain types of storage heater may contain fans to assist in distributing the warm air and improving circulation. This will cause better convection of the air within the room space as warm air is blown from the heater and cool air is drawn in to take its place.

Embedding electric heating elements into the floor, walls or ceiling of a room provides an alternative source of electric heating. This form of heating has, however, been the subject of much adverse publicity in the 1950s and 1960s in relation to high running costs. Nationally inefficient methods of electricity production are held largely responsible for the expense of electrical energy. Power stations are often regarded as only 30–35 per cent efficient in their production of electricity although production by nuclear methods claims greater percentage efficiencies.

DUCTED WARM AIR[5]

Electricity, gas or solar energy may be the heating source used in ducted warm air heating. The systems consist of metal ducts through which the heated air is fan-assisted to the outlet grilles in the various rooms. Figure 5.10 shows a typical domestic installation.

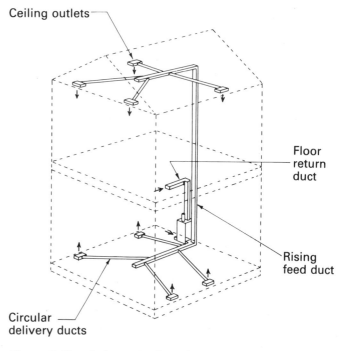

Figure 5.10   A domestic ducted warm air heating installation

Factors to be considered in warm air heating are:

(a) The volumetric capacity of the spaces served, and the implications of this on the rating capacity of the heat source.
(b) The fall in temperature experienced before the air is discharged into the various room spaces. Air is a poor conductor of heat and is associated with rapid temperature falls because of its inability to hold heat.

Ducted warm air is the principal method employed for heating dwellings in North America. Systems, particularly gas fired, are available in an extensive range of capacities, but all systems rely on the same basic concept of fan-assisted delivery of warmed air to individual rooms. Many of the arrangements employ a duct layout as illustrated in figure 5.10, where delivery into the rooms is via perimeter-located diffusers. An alternative to this layout may be the 'short-duct' system which tends to be used in well-insulated properties. This method provides for delivery of warmed air from the centre rather than the perimeter of the room.

It is claimed that the air circulation achieved with warm air systems is so efficient that there is likely to be less than 2°C difference between the air temperature at ceiling level and the air temperature at floor level. Since there is also the facility to recycle warmed air, any advantitious solar heat gains in certain rooms will be distributed throughout the whole of the premises. This means that rooms not currently receiving direct solar radiation will also benefit.

Diffusion of the warm air into the roomspace will be achieved most commonly using floor perimeter diffusers, ceiling perimeter diffusers or side wall registers. Floor perimeter diffusers should be located a short distance away from the wall to prevent the obstruction to air movement which could be caused by curtains. Ceiling perimeter diffusers should not be used in rooms with floor-to-ceiling heights in excess of 2.5 metres or in rooms that are not heated at floor level. Side wall registers are installed above skirting level where their air flow will be directed towards the windows. High-level mounted wall registers are used in conjunction with either low-level wall registers or perimeter floor diffusers. The recycling of air which is a feature of ducted warm air systems will only occur by extracting from certain rooms. Rooms such as bathrooms which will generate high quantities of water vapour are excluded, as are those rooms which could cause the circulation of fumes of undesirable odours (kitchens and WCs).

SOLAR HEATING

Considerable research has been undertaken in recent years relating to the viability of solar energy for space heating and for heating hot water (see chapter 2). Before examining the basic component parts of a solar heating system, this section begins with consideration of the influence of glazing on internal heat gains.

Glazed areas have been the subject of much deliberation in earlier years and particularly in the 1980s in respect of heat losses. The amendments to the *Building Regulations* issued in 1982 and subsequently embodied into Part L of the *Building Regulations 1985* have completely revised the performance needs of buildings in terms of thermal insulation. In achieving a *U* value of 0.6 W/m² K for walls, special consideration is given to the glazing. The

regulations now limit the maximum permissible areas of glass on the perimeter face of the building and on its roof.

In performance terms, single glazing is regarded as having a *U* value of 5.7, double glazing 2.8 and triple glazing 2.0. The loss of internally generated heat through glass is clearly of concern but it should also be recognised that the 'greenhouse effect' in relation to incident solar heat may be a useful facet of glazing. This effect arises as a change in wavelength occurs when short-wave penetrating rays change to long-wave radiation after passing through the glass. The change in wavelength discourages heat loss back through the glass, causing the internal temperature to rise.

In the study of solar heat gain through glazing, clearly the orientation of the building is an important factor (the way the building faces relative to the compass and, in particular, the way in which the building faces relative to the path of the sun). Accumulation of glazing on the south-facing wall will have a marked effect on internal heat gain when compared with equal distribution of the glazing on all elevations of the building. Many of the solar control glasses produced by Pilkington Brothers (St. Helens) are designed for exclusion of solar light and solar heat. Pilkington Brothers also produce energy conservation glass (Kappafloat glass) which is specifically designed to enhance the greenhouse effect and also reduce the rate of loss of internally generated heat.

Chapter 2 has already considered the use of solar energy to produce domestic hot water, this section outlines the use of solar energy in space heating.

As stated in chapter 2, the climate of the UK is unlikely to support fully the all-year-round needs of hot water for the building, and this statement also applies to space heating. Since hot water is used throughout the year and yet space heating is only necessary for part of the year, this makes the viability of solar space heating even more questionable.

The fundamental component of a system for gathering solar energy is the 'collector'.[6] Various materials (largely metals) may be used for the collector, its function being to absorb heat for transfer into the building by either air or water. If the transfer medium is air, natural convection is a common means of air movement (figure 5.11).

Where water is used to collect solar heat, storage is a common feature. Storage may be provided by a water tank or by pebbles over which the water is routed. In water collection systems a light sensor will be used to activate the pump in the arrangement, thereby ensuring

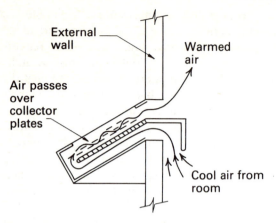

Figure 5.11   Solar air heating

that water is passing through the collector only when reasonable levels of solar radiation are being received. Figure 5.12 shows the component parts of a solar water collection system.

GROUP AND DISTRICT HEATING

Both group and district heating schemes involve the use of the generation of heat at a heat station remote from the buildings being served. In group heating the supply will be to a number of similar buildings, while in district heating a variety of buildings may be served.

The heat station is ideally situated centrally in the development to regularise the temperature of delivery.

Examples of such systems are not common in the UK but systems are used extensively in the USA.

Under the category 'group/district heating' there are a number of classifications of systems:

(a)   the two pipe system;
(b)   the three pipe system; and
(c)   the four pipe system.

In the two pipe system there is one pipe for the flow and one for the return, and connections are made to these for both space heating circuits and indirect hot water for sanitation.

Three pipe systems recognise the difference in heating demands with season. There are two flow pipes, one large for winter use and one small for the summer when only sanitation hot water is required. Figure 5.13 shows a typical three pipe layout.

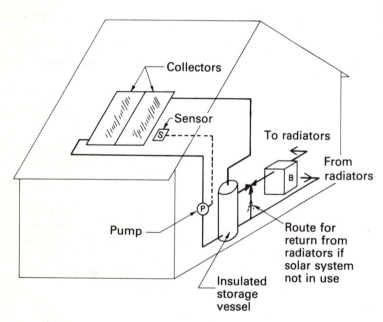

Figure 5.12   Solar water heating

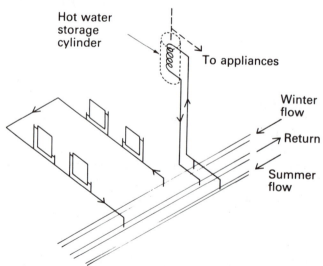

Figure 5.13   The three pipe group heating arrangement

Four pipe systems have a flow and return pipe for both the heating circuits and the sanitation hot water. It should be noted that hot water for sanitation is not stored in this detail but taken direct from the main supply. Figure 5.14 illustrates the four pipe layout.

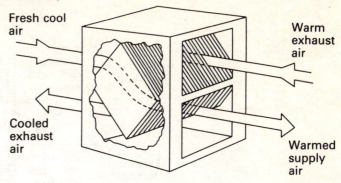

Plate heat exchanger

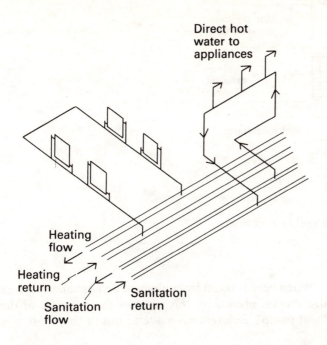

Figure 5.14   The four pipe group heating arrangement

Advantages of communal heating and hot water supply include:

(a)   The overall fuel consumption of the buildings in the group or district tends to be lower than that which would have been consumed by individual users if a conventional heating and hot water system had been used.
(b)   The heat source may be a by-product of electricity generation (waste gases from the turbines), thereby increasing the efficiency of electricity generation.
(c)   Increasingly, incinerated refuse is used to provide the heat energy for communal heating.
(d)   The use of a single heat station should provide an overall saving in equipment and attendance costs (despite the quantity of distribution pipework).

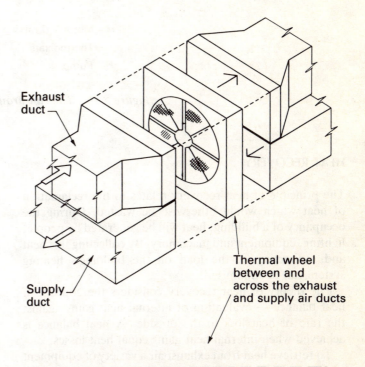

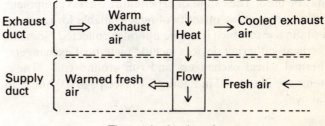

Thermal wheel exchanger

Figure 5.15   Plate and thermal wheel heat exchangers

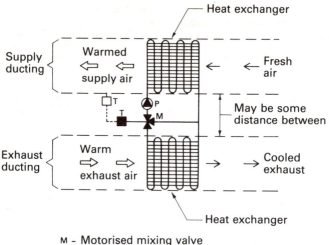

Figure 5.16   The run around coil heat exchanger

HEAT RECOVERY

The principle of heat recovery relates to the reclamation of heat which would otherwise be wasted. During the occupancy of a building, heat will be generated by people, lighting, equipment and machinery. By collecting this heat and redirecting it, the load on the building's heating system can be minimised.

The theory of heat recovery considers the concept of heat balance — evaluation of internal heat gains against the rate of heat loss to the outside. A heat balance is achieved when internal heat gains equal heat losses.

To retrieve heat from exhaust air a variety of equipment may be used. The extent of expenditure on this equipment will depend on the quantity of heat available for recovery and the use to which it can be put once collected. Some of the heat-gathering devices include plate heat exchangers, thermal wheel exchangers and run around coils. These items are illustrated in figures 5.15 (see page 43) and 5.16 (above).

When heat is taken by equipment and enhanced before distribution elsewhere, this involves the principle of the 'heat pump'. Reference is made to this in chapter 6.

REFERENCES

1 *CP 341.300–307: 1956 Central heating by low pressure hot water.*
2 *CP 3000: 1955 Installation and maintenance of under-feed stokers.*
3 *The Gas Safety Regulations 1972*; Factory Acts; Clean Air Acts 1956, 1968.
4 Chartered Institution of Building Services (CIBS), *IHVE Guide, B13 Combustion Systems.*
5 *BS 5864: 1980 Code of practice for installation of gas-fired ducted-air heaters of rated input not exceeding 60 kW.*
6 British Standards Institution, *Drafts for Development, DD: 1982 Methods of test for the thermal performance of solar collectors.*

6 SPECIALIST SERVICES — MECHANICAL VENTILATION, AIR CONDITIONING, LIFT INSTALLATIONS, ESCALATORS AND FIRE FIGHTING INSTALLATIONS

MECHANICAL VENTILATION

Introduction

The interiors of buildings are ventilated in order to provide environmental conditions in which the occupants may operate with comfort, safety and efficiency. By the provision of ventilation we may ensure that:

(a) Sufficient fresh air is available for respiration — the amount consumed by the occupants will be influenced by their natural metabolic rate, and this in turn is determined by the characteristics of the individuals and their rate of activity.

(b) Provision is made for the extraction of pollutants which may accumulate

      (i) naturally (for example, carbon dioxide),
      (ii) through the processes of occupancy (for example, tobacco smoke), or
      (iii) through the production processes undertaken by the occupants (for example, fumes from industrial processes).

(c) The level of internal relative humidity can be controlled.

(d) A suitable environment is created with respect to air temperature.

(e) Air is available to support the processes of combustion (for example, to support complete burning of gas fuel at a domestic gas fire).

(f) Fumes and smoke arising from accidental fires can be controlled.

The geometry of large buildings and, in particular, their plan shape will prevent complete ventilation of the enclosure space through perimeter fenestration. The physical distance between central areas and the windows will mean that air stagnation is most likely to occur in the central areas of the building, and to relieve this situation mechanical ventilation can be employed.

When mechanical ventilation is applied to buildings, the systems used may be:

(i) systems which deliver air which is unheated;
(ii) plenum systems which heat the air for buildings which have no other heating system (figure 6.1);
(iii) tempered air systems in which the air is heated for buildings which have also a separate heating installation.

When considering the deployment of natural or mechanical ventilation to a particular premises, the criteria considered in respect to the air are generally quality, quantity and controllability. If natural ventilation is contemplated, the following limitations should be appreciated:

(i) precise control of air supplied by natural ventilation is not possible, and
(ii) the occupier of a building can introduce a variable into the provision of fresh air supplies by opening and closing windows in an unpredictable way.

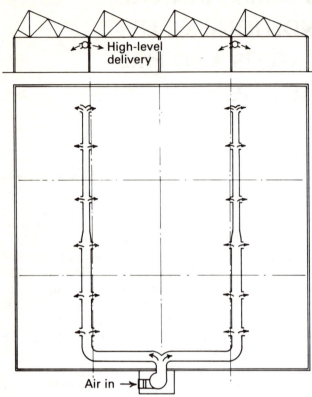

Figure 6.1   Plenum ventilation for a factory building

Mechanical ventilation will be necessary:

(a)   where spaces within the building cannot be naturally ventilated;
(b)   in premises where it is necessary to remove pollutants (such as fumes or dust);
(c)   in special premises such as hospitals where a sterile environment is required.

The use of mechanical ventilation will allow provision to be made for the control of air temperature, humidity and purity, in addition to helping control air movement.

Buildings of different use will demand different quantities of fresh air for their occupants. This may be expressed in terms of litres/s (dm^3/s), or in terms of room air changes per hour. A key factor in the assessment of the extent of air required within rooms is often the extent of tobacco smoke produced. The CIBS guide[1] suggests that the quantity of fresh air needed can vary from 8 litres per person per second in fairly smoke free environments, to as much as 25 litres per person per second where heavy smoking is anticipated.

The method by which air is brought into a building gives rise to a classification of the type of system which may be provided:

(i)   natural inlet — natural outlet;
(ii)   natural inlet — mechanical outlet;
(iii) mechanical inlet — natural outlet; and
(iv)   mechanical inlet — mechanical outlet.

Figure 6.2 illustrates this classification.

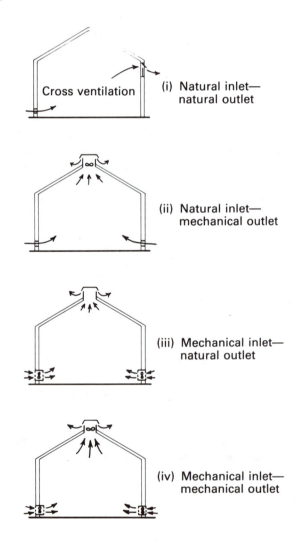

Figure 6.2   Classification of ventilation systems

With natural entry of air and mechanical extraction, the movements of the air within the building may be influenced by cross ventilation, or by the chimney (stack) effect, experienced in tall buildings by the rise of air through features such as stairwells. Here the cooler air found in the stairwells encourages the movement of warm air from the surrounding rooms into the well, where the warm air naturally rises creating a chimney effect.

Mechanical exhaust systems may give rise to uncomfortable conditions caused by draughts if the rate of expulsion results in suction of air into the room spaces. In contrast, mechanical inlet and natural outlet systems can cause small positive pressures within rooms which will discourage air ingress and thereby prevent draughts.

Mechanical inlet and mechanical outlet arrangements should provide the greatest degree of control in that both the rate of input and the rate of extract can be carefully monitored. The *Building Regulations 1985* make specific stipulations with regard to the rate of air change in certain room areas. These areas include habitable rooms, kitchens and bathrooms of dwellings, common spaces in buildings containing dwellings, and sanitary accommodation in any building. The air changes per hour recommended by the regulations range from 1 per hour for habitable rooms to 3 per hour for kitchens, bathrooms and sanitary accommodation spaces.

Fans

In ensuring the correct rate of delivery of air to or from different spaces within a building, electric fans are employed. There are many arrangements of fan possible but few are sufficiently efficient for the movement of air in service installations. Propeller fans, for example, are less than 40 per cent efficient. Those suited to air movement include:

(a) centrifugal types (analogy — playground round-about);
(b) axial flow (analogy — the jet engine turbine);
(c) tangential flow (analogy — the paddle steamer).

Centrifugal fans are the most widely used, with efficiencies of around 85 per cent. The best axial flow details are approximately 75 per cent efficient, and the tangential flow types around 50 per cent.

Ducts for Air

When designing ducts for the conveyance of air, a number of features should be incorporated:

(i) The duct should be smooth — this will not only help the uniformity of air movement but also reduce the possibility of noise generation by turbulence as air moves over obstructions.
(ii) The number of changes of direction of the duct should be minimised — for circular ducts the formation of bends should be achieved gradually with as large a radius as practicable; for rectangular ducts 90° changes of direction may be 'smoothed' by the incorporation of duct vanes (figure 6.3).
(iii) Branches off main ducts should be made with a gradual curve and not with a sharp 90° connection.

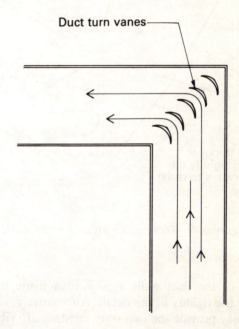

Figure 6.3   The use of duct vanes at corners

Galvanised mild steel is probably the most widely used duct material. This may be pressed into rectangular sections with riveted or bolted joints, or helically wrapped flat sheets to produce circular ducts. As the size of the duct increases there is a greater possibility of the wall vibrating (drumming) as a result of the air movement, and

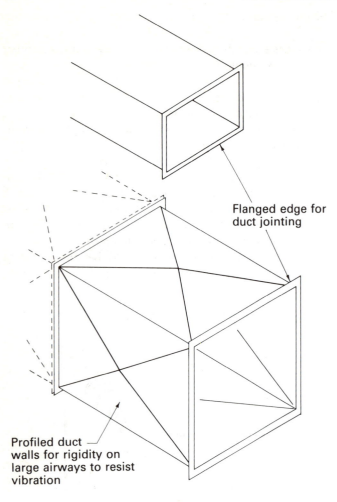

Flanged edge for
duct jointing

Profiled duct
walls for rigidity on
large airways to resist
vibration

Figure 6.4   Rectangular pressed metal ducts

shaping of the duct walls as shown in figure 6.4 may increase the rigidity of the detail. Alternatively, attached braces may provide the necessary resistance to vibration.

In addition to galvanised mild steel sheet, a number of other materials can be used for ducts including PVC, asbestos–cement, and various metal alloys. When materials other than galvanised mild steel are used, this is often for specialist installations where resistance to the corrosive properties of fumes or sealed air-tight joints are necessary. Asbestos–cement is commonly found in situations where its use has been restricted to exhaust systems.

Efficient delivery of air to room spaces through ducts will involve the detailed calculation of desired air flow velocities which will not only satisfy the volumetric needs of the room but will reflect the frictional losses experienced during conveyance through the ductway.

Towards the points of delivery, the cross-sectional profile of the duct may become progressively smaller as a smaller volume of air is needed. Such changes in duct section should occur gradually if the velocity reduction caused by friction is to be minimised.

Grilles, Diffusers and Dampers

At the point of air delivery or extraction the duct is fitted with a grille or diffuser. The type of fitting used will have a considerable influence on the spread achieved in distributing the air into the space. Conical diffusers are best for efficient 'throw' of air into the space and can take delivery velocities of around three times those discharged through simple slot diffusers without generating excessive noise (figure 6.5).

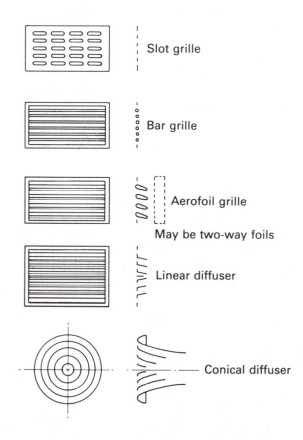

Slot grille

Bar grille

Aerofoil grille

May be two-way foils

Linear diffuser

Conical diffuser

Figure 6.5   Air delivery grilles and diffusers

Dampers within the ductwork control the flow of air by varying the resistance to air flow, and these may be manually or automatically operated. Special dampers will be needed to resist fire spread, and these will generally be positioned whenever the duct passes through elements of the building (walls and floors) which have been designated as fire compartments. Figure 6.6 shows some typical dampers including one for the containment of fire. Fire details may be of steel plate which will provide 2 hours' resistance and these will be activated on the failure of fusible links at approximately 70°C.

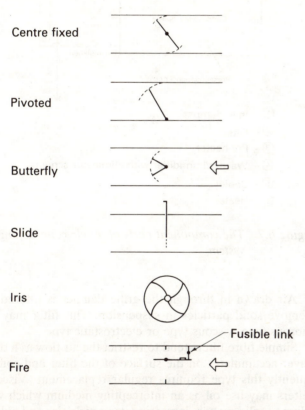

Figure 6.6   Duct dampers

Noise Generation and Transfer

As ventilation ducts may extend to all areas of a building they provide potential routes for noise transfer, and this possibility must receive special consideration from the services engineer. The sources of noise include:

(a)   that generated in service plant rooms by service machinery;

(b)   that caused by vibrating duct walls;

(c)   that generated as a consequence of the moving air passing over or through obstructions (for example, dampers);

(d)   that resulting from air leaks in the ducting;

(e)   that generated within the various areas of the building by the occupants or their machinery.

Service machinery is generally provided with special impact-absorbing mountings to reduce the possibility of flanking transmission of sound. Any ventilation or air conditioning system has potential problems of noise generation and transfer caused by the incorporation of fans for air movement.

AIR CONDITIONING

Introduction

As the name suggests, true air conditioning involves conditioning of the air, not simply heating or cooling. Conditioning of the air will mean purification and the control of relative humidity.[1]

As a consequence of its molecular structure, air can hold varying amounts of water vapour according to its temperature. The quantity of water vapour in a given quantity of air may be expressed as a percentage of the total quantity of vapour capable of being held by the air at that temperature, this being termed the *relative humidity*.

Extreme high or low levels of relative humidity can cause discomfort, but levels between 40 and 60 per cent should create a reasonable internal environment.[1]

At dew point temperature,[2] condensate will begin to be precipitated from the air on to the surface of the enclosure surfaces in a surface condensation effect, or within the body of the enclosure materials as interstitial condensation. By controlling air temperature and relative humidity through air conditioning, the occurrence of condensation can be minimised. Provision of an air conditioning system should ensure a regulated supply of air for breathing, the removal of the products of respiration, and removal of the by-products of occupancy. Approximately 1 litre per second per person of fresh air is needed to ensure that the levels of carbon dioxide generated by breathing do not cause discomfort. Air conditioning will ensure that this demand is met but care needs to be exercised to prevent nuisance through the adoption of a conditioning system. The incoming air velocity, direction of supply and noise are all potential problems to be kept in mind. In buildings where air conditioning is to be used, the Architect and Services Engineer should liaise for successful integration of the system into the design. Correct siting of air inlets and exhaust outlets will help prevent nuisance in operation and increase the efficiency of air delivery.

Air as a material is less predictable than a liquid, with a number of factors influencing its unpredictability:

(a)   the extent of natural ventilation;
(b)   the velocity of mechanical ventilation;
(c)   natural convection;
(d)   the geometry of the room and its obstructions;
(e)   occupant movements within the room.

Natural advantitious ventilation will always occur through windows and doors but the calculation of the extent of this will be dependent on unpredictable variables such as the strength of the incident wind. The air velocity of mechanical air inputs will depend on the type of fan used and the pressure losses incurred as the air flows through the delivery ductwork. Broad classifications of delivery velocities include:

(a)   low-pressure systems — velocities of up to 10 metres per second; and
(b)   high-pressure systems — velocities of over 10 and up to 20 metres per second.

Velocities in excess of 20 metres per second are prone to exaggerated development of noise and require special attention in design.

Air Conditioning Systems[3]

There are many variations possible in the arrangement of the components of air conditioning systems. Some designs are for summer use only and others for all year use. Some systems have the facility for reclamation of air from the room space, others simply have an exhaust extractor without recirculation. Figure 6.7 shows the typical component parts of a system. In this diagram it should be appreciated that the controls, sensors and valves have been omitted for clarity of presentation.

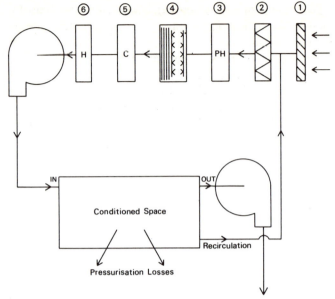

① – Inlet damper
② – Filter
③ – Pre-heater
④ – Washer/humidifier with eliminator screen
⑤ – Cooler
⑥ – Heater

Figure 6.7   The component parts of an air conditioning system

Air drawn in through the grille damper is filtered to remove solid particles in suspension. This filter may be fibrous type, viscous type or electrostatic type.

Simple fibre filters tend to restrict the air flow as a dust layer accumulates on the surface of the filter and consequently this type requires regular replacement. Viscous filters may use oil as an intercepting medium which will cling to solid matter, increasing its particle size and aiding its collection. Electrostatic filters rely on the electrical-potential created between conductive metal plates to cause solid matter to become charged and drawn towards the conductive surfaces of the plates.

Pre-heating may aid the achievement of the desired level of relative humidity by later treatments in the system.

Washers may be water or steam, the former atomised to help absorption by the air. The washer will provide the means for humidification of the air but it also acts as a filter of matter which has escaped the earlier filter. The washer is often used in conjunction with eliminator plates as implied by figure 6.1, these performing the role of a crude filter for water droplets created by the intense washing process.

Air is usually heated by passing it across a hot water pipe coil exchanger which is also the same basic apparatus used for cooling the air. When cooling the flow, chilled water enters the coils from a refrigerator; chilled brine (salt water) is a common coolant used in cooling coils.

A simple refrigeration cycle is illustrated in figure 6.8. This shows the four basic component parts of a refrigerator: the evaporator, the compressor, the condenser and the expansion valve.

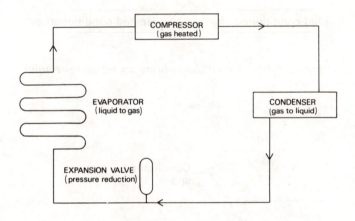

Figure 6.8   The basic components of a refrigerator

Refrigerators use special refrigerant liquids which have a particular ability to absorb and lose heat.[4] One such refrigerant is dichlorodifluoromethane or R12.

At the evaporator the liquid refrigerant absorbs heat and evaporates, and this gas is then compressed causing the temperature to rise considerably. On reaching the condenser, the refrigerant loses heat and condenses back to a liquid. Still under high pressure, it enters an expansion valve which reduces the pressure of the liquid before it enters the evaporator to continue the cycle.

Heat Pumps

Since a refrigerator absorbs heat into the refrigerant at the evaporator and emits heat from the condenser, it may be used to perform the function of a heat pump. Heat generated at the condenser may be used to heat a building by using the refrigeration cycle to absorb heat from a source, enhance this heat by compression, and then deliver this into the room spaces with fan assistance.

Heat may be absorbed into the evaporator in a low-grade, low-temperature form (for example, from a water course), compressed and then dissipated from the condenser in a higher-temperature form.

Although the temperature of the heat source is low, in effect it is naturally available heat from which, through the application of the heat pump, some function has been derived. The value of the heat obtained from the condenser is determined by adding the heat absorbed by the evaporator to the heat generated by compression. Should the evaporator absorb 5 units of heat and the compression cycle generate a further 1 unit of heat, the total heat from the condenser would be 6 units. Since one unit of heat has been consumed in the production of a total of six units, this is termed as an 'advantage' of 6:1. 'Advantage' is a reflection of the performance of the heat pump.

Dual Duct Systems

Figure 6.9 shows the route taken by conditioned air on its way to the room space and indicates the normal practice of having the cooler and heater in line. A variation of this arrangement is provided by the dual duct system where the conditioned air is cooled and heated in separate ducts (figure 6.9).

By separating the heating and cooling equipment, better control can be effected over the system. The inlet to the room will generally be provided by a constant-velocity mixer which blends the cool and warm air together to meet the needs of the room. A variable damper, located at the point where the cool and warm air are brought together, will be operated by a room sensor (figure 6.10).

Induction Systems

The system of induction air conditioning may also provide flexibility to meet the demands of a room space by separating the heating and cooling from the ducted conditioned air supply. Figure 6.11 illustrates such a system where perimeter convection outlets mix conditioned air with convected warmed or cooled air. In summer months the convected air entering the unit will pass through cooling coils and in winter through heating coils.

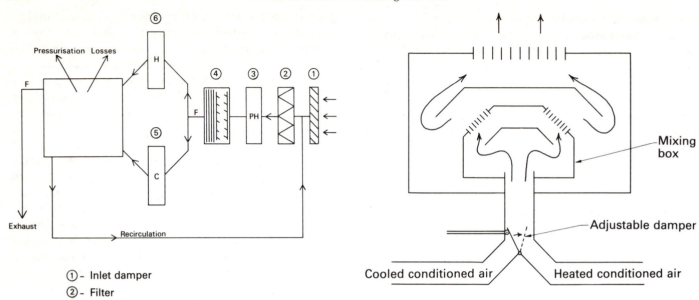

① - Inlet damper
② - Filter
③ - Pre-heater
④ - Washer/humidifier with eliminator screen
⑤ - Cooler
⑥ - Heater
F - Fan

Figure 6.9   The dual-duct air conditioning system

Figure 6.10   A constant-volume air mixer inlet unit

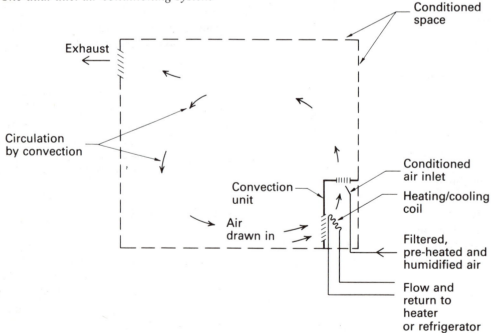

Figure 6.11   An induction air conditioning unit

Independent Air Conditioning Units

Air conditioning may be provided by a central air conditioning plant with conditioned air ducted to and from the room spaces or alternatively individual air conditioning units may be provided to each room. These contain the basic components of the air conditioning plant in miniature. The room air is drawn in at the conditioning unit for explusion, and fresh air drawn from the external air. This type of unit is used extensively in regions with hotter climates than that of the UK, in particular the USA.

Variable Air Volume Air Conditioning

Energy conservation is of particular importance because of the high costs of fuel, and consequently any measure to reduce energy consumption will prove attractive to the building occupier. Variable-volume air conditioning provides a means by which energy consumption can be more efficiently linked to the demands of the air conditioned spaces within buildings. Sensors in the room are used to regulate the supply to suit the demands of that room, and delivery at a uniform rate to each area is therefore prevented. The cumulative effect of each room only taking in conditioned air to suit its needs results in an overall lowering of the energy consumed by the system. Costly control sensors and equipment are demanded by this system but the initial outlay is rapidly compensated by improvements in efficiency which will ensure long-term savings.

LIFT INSTALLATIONS

Introduction

The incorporation of lift installations into a building, as with other specialist services, will ordinarily necessitate liaison between the building designer and the service manufacturer. The installation will follow the integration of design criteria generated by both of these parties.

The designer determines the number of floors to be served and, through the design, the types and quantity of users. Functional demands will be constraining in terms of user requirements and in terms of quality standards. It should be remembered that quality as a design parameter will not only relate to the overall prestige of the building

but to practical considerations such as the accuracy of levelling at each floor landing.

From the manufacturer's point of view there will be the need to accommodate basic components which are fundamental to the type of lift chosen, together with the need to ensure safety in operation.

A useful reference which summarises the demands made on both the designer and manufacturer is BS 5655: 1985.[5]

Traditional electric lifts will require machine rooms of size to suit the sophistication and extent of the installation. The usual location for this feature is either at the top of the lift shaft or at the bottom of the lift shaft, depending on the roping arrangement employed for the lift car. Figure 6.12 shows some of the typical parts of the shaft including the machine room. Note the provisions made for ventilation.

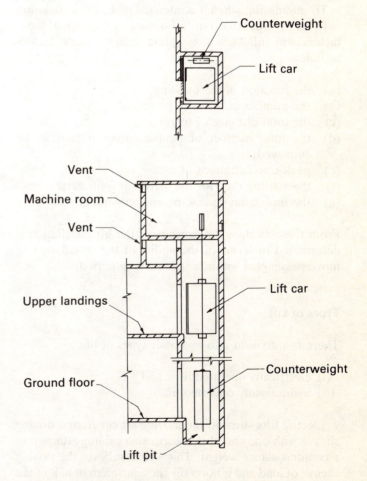

Figure 6.12   A typical lift shaft

The location of the lift shaft will depend on the location of the entrance to the building and the design of the circulation spaces as attempts are made to optimise the efficiency of pedestrian flow. In larger buildings a conflict may arise in design between the economic benefits of grouping together lift shafts and the degree of efficiency of pedestrian flow that this causes.

The speed of the lift is a design feature which again will relate to the function of the building. In shops and stores speed of circulation may be of particular importance, while in residential buildings, reliability may be a more important criterion than speed of operation. Speeds of 0.5–1 metre per second are common, although speeds in excess of 5 metres per second are possible. High speeds tend to relate to high costs, and may only be justifiable in very tall buildings where acceleration between floors could be a significant factor in efficiency.

To summarise, when it is intended that a lift installation is to be incorporated into a building, a number of basic factors will influence the system design. These factors include:

(a) the function of the building;
(b) the number of floors to be served;
(c) the total distance of travel;
(d) the total number of people and/or objects to be conveyed;
(e) peak concentrations of use;
(f) the waiting time for the arrival of a lift car;
(g) the time delay in loading and unloading.

From these factors, the capacity of the lift installation is determined in terms of the ability of the installation to move passengers within a specific time period.

Types of Lift

There is a division into two basic types of lift:

(i) electrically operated lifts, and
(ii) hydraulically operated lifts.

Electric lifts suspend the moving lift car from a driving sheave with one end of the suspension cabling attached to a counterbalance weight. This weight relieves the driving sheave of load and is normally the equivalent of half of the weight of the loaded car.

The principle of a pulley system may be employed to reduce the load on the driving mechanism further, as shown in figure 6.13. This shows both a single wrap arrangement for lightly loaded cars and a 3:1 wrap for more heavily loaded arrangements.

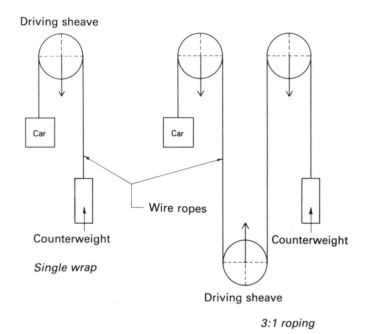

Figure 6.13　*Single wrap and 3:1 lift roping*

When a single wrap roping suspension is used the machine room will be at the top of the lift shaft, and where the 3:1 roping is used the machine room will be at the bottom of the shaft. The single wrap provides a compact installation and small shaft dimensions, such as are shown in figure 6.14.

Safety[6] in the use of a lift system is ensured by two components of the installation: the overspeed governor and the safety gears. Safety gears are effectively brakes which come into operation in response to a reaction from the overspeed governor. The governor is connected to the car by a rope and pulley and, in the event of the achievement of a speed beyond a pre-set limit, the governor will automatically activate the brakes.

Hydraulic lifts[7] operate in a fundamentally different way from the electric lift and, as a generalisation, require far fewer operational controls.

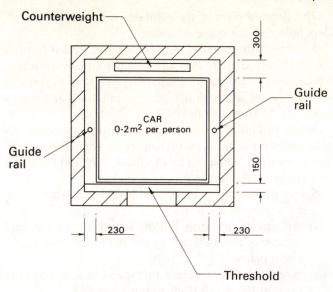

Figure 6.14   A compact lift shaft detail

Oil supplies the driving force to cause movements of the car by the use of a driving ram. The principle of operation may be compared to the operation of a hydraulic jack used on motor cars. A rise of the ram is caused by pressuring the oil, and a fall is achieved by releasing the drive pressure.

Within the range of hydraulic lifts available two positions are found for the location of the drive ram, either directly beneath the lift car or attached to the side of the lift car. Structural loads may only need to be carried by one wall of the lift shaft where certain hydraulic lifts are used, and this may prove a considerable advantage in design. Major savings may also occur by comparison with the electric lift in that a high-level machine room is not required. The control equipment for a hydraulic lift may even be located in a position remote from the lift shaft.

The hydraulic lift is known for its smooth ride and for accurate levelling, although its working height is often limited to around 30 m.

Lift Call Systems

Apart from the efficiency limitations imposed by the type of lift chosen, the effectiveness of the installation can be considerably influenced by the form of call system used.[8] The options for call systems include:

(a) collective control — down collective, or directional collective
(b) automatic control
(c) group control, and
(d) attendant control.

Down collective — here the car reacts to a call and ignores all other calls while in upward motion. On the way down, however, it will call at all floors where a call has been made.

Directional collective — while in upward motion the car will only call at successive floors which have requested upward movement. The calls for downward movement are recorded and are met in a progressive downward motion during which upward calls are recorded but ignored until the lowest level is reached.

Automatic control — this is a simple form of call system in which no record of calls is made while the car is in motion. The car will only respond to a call after having become stationary.

Group control — in large installations where a number of cars are used, a computer is used to mastermind car movements to ensure efficiency of movements.

Attendant control — here the car stops by the operation of the attendant's control which is activated just before reaching a desired stop level.

ESCALATORS

Introduction

The first operative escalator was used at the Paris Exposition in 1900. Since then, escalators have been installed in situations where rapid movement of large numbers of people is needed and applications have proved most useful in airports, underground stations and large retail stores.

Direct comparisons with lift installations are difficult because of the different methods of operation; escalators deal with continuous loads while batch loads are conveyed by a lift.

BS 5656: 1983 concerns the safety rules for the construction and installation of escalators and passenger conveyors. In addition to safety, other design features include:

(a)   a high carrying capacity;
(b)   no waiting interval;
(c)   flexibility (since reversible);
(d)   reliability.

Typical dimensions may allow for up to 10,000 persons per hour capacity where the conveyance speed is around 0.5 metres per second. Figure 6.15 shows some typical dimensions and indicates the space necessary to accommodate the component parts. The need for clear headroom space is also featured.

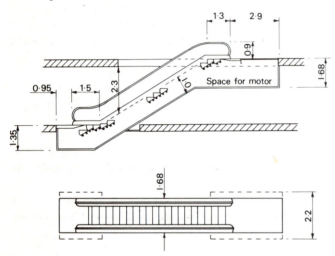

Figure 6.15   Escalators — typical installation dimensions

The slope or pitch of the escalator is usually 30° or 35°, depending on the space available.

Reversible direction of motion provides added flexibility which can be put to best advantage in situations such as underground stations to allow for peak flow movement preferences. The London Underground and Paris Metro both benefit from this flexibility.

When installing the escalator, the most common arrangement is the criss-cross layout as shown in figure 6.16. The dimension *D* marked on this illustration can be varied to suit the available space.

Safety of passengers is ensured by the following:

(i)   An emergency stop button found at the top and bottom of the escalator by the potentially dangerous comb plates.
(ii)   A governor to regulate the speed, and specifically prevent the detail from running too fast.
(iii) Automatic controls to bring the detail smoothly to a halt in the event of a power or mechanical problem.
(iv)   A handrail that moves at exactly the same speed as the steps.
(v)   Balustrades, handrail and side panels which discourage the catching of clothing or packages.
(vi)   A level of illumination of at least 100 lux to the steps.

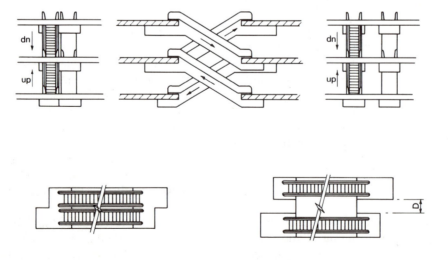

Figure 6.16   Escalators serving successive floors

Many of the buildings served by escalators are compartmentalised for the purpose of fire confinement. To preserve the integrity of such fire compartments, the escalator can be fitted with steel roller shutters which will close off the stairway on the activation of a smoke detector, heat detector or fusible link (figure 6.17).

Sprinklers may provide an alternative source of fire control. These may be used in conjunction with fans which can be utilised to pressurise the stairwell by a system of 'spray nozzle curtaining'. Here, high-pressure water nozzles create a water curtain to discourage flame and smoke movement, these being activated either by heat or smoke detectors.

In contrast to escalators, passenger conveyors have a very shallow incline, so eliminating the need for steps. The result is a continuous conveyor belt which can carry passengers and luggage. With operation speeds at more than twice those of the escalator, around 1–1.3 metres per second, the capacity to carry passengers is very high.

It is particularly important for the passenger conveyor to have sufficient space available at the point of termination to allow free movement for those leaving it, so minimising congestion at this point.

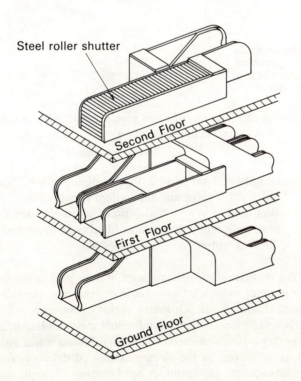

Figure 6.17　Escalators — the use of fire roller shutters

FIRE FIGHTING INSTALLATIONS

Introduction

When details are incorporated into a building specifically with fire control[9] in mind, two classifications of control may be made:

(i) Passive control — resulting from the nature of the materials of construction. Incombustible and low combustion materials built into the structure will discourage fire spread and in so doing exercise a passive control over fire development.
(ii) Active control — resulting from inclusion of equipment specifically for fighting an outbreak of fire.

Passive control in-built into the structure will be enhanced by the need to comply with the *Building Regulations*.[10] These regulations dictate the extent of fire resistance to be provided by the elements of the structure, and features such as the rate of surface fire spread. Large buildings may have to be compartmentalised, with specific fire resistance qualities to be achieved by the enclosure elements of the compartment.

The items of equipment which are used to provide active control in fire fighting may be grouped under the headings 'portable' and 'fixed'. Examples of portable equipment include extinguishers,[11] blankets,[12] fire buckets and tools such as axes.

Extinguishers are generally between 4.5 and 9 litres capacity and one of a variety of types:

(a) carbon dioxide;
(b) water — gas pressured (cartridge inside)
　　　— soda/acid
　　　— stored pressure (compressed gas above the water);
(c) dry powder;
(d) foam; and
(e) vaporising liquid.

It is particularly important that the appropriate extinguisher is provided to fight the anticipated form of fire. The table which follows summarises the application of different extinguisher types. It should be noted that although suitability may be indicated in the table, special variations of the selected extinguisher may be required in particular circumstances.

Extinguisher type	Form of fire			
	Electric present	Solids, such as paper and wood	Flammable liquids	Gases
Carbon dioxide	Yes	Yes	Yes	No
Water	No	Yes	No	Yes
Dry powder	Yes	Yes	Yes	No
Foam	No	Yes	Yes	No
Vaporising liquid	Yes	Yes	Yes	No

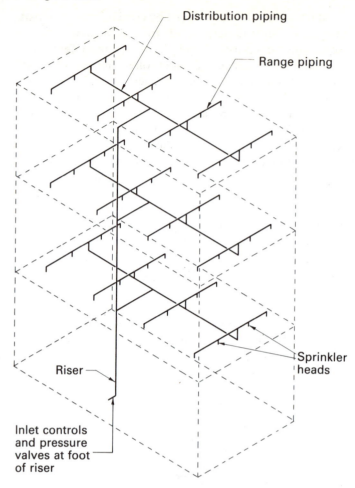

Figure 6.18   The distribution pipe network for a typical sprinkler installation

Permanent equipment, in contrast, may include:

> (i)   sprinkler systems;[13]
> (ii)   permanent hoses;[14]
> (iii)   riser systems — water or foam.[15]

The use of sprinklers is usually associated with certain types of buildings in which the performance relates to fire containment and the protection of goods. Commercial, retail and industrial premises represent the main groups of buildings provided with this feature of control. Installation also generally has the beneficial effect of lowering the building insurance premium.

Within the arrangements of sprinkler available there are three basic systems:

(a)   Wet systems — these consist of a pipe network, typically as illustrated in figure 6.18, which is always filled with water. When the sensor located at the sprinkler head activates the system, water immediately falls to the area below the outlet. This type of system may not be suitable for use in unheated buildings where the danger of pipe freezing exists in winter months.

(b)   Dry systems — these overcome the problems of use in unheated buildings since the distribution and range pipes are filled with air until a sprinkler head is activated. On the outbreak of a fire the release of the compressed air stored in the piping allows water entry via a valve and the system rapidly fills.

(c)   Wet and dry systems — these allow for seasonal variations and will be water filled in summer and air filled in winter.

An additional classification of system may arise from the manner in which the water is delivered to the fire by the sprinkler head. Ordinary sprinkler systems will be activated by heat action on individual sprinkler heads, and only those heads affected will open allowing water entry. By comparison, in the deluge system of delivery, room sensors operate the sprinkler head causing all the heads in a particular area to open at once to smother the fire.

The catchment area over which a sprinkler head discharges will be determined by the water delivery pressure and the design of the outlet head. Coverage is generally between 7.5 m² in high risk buildings and 21 m² in low risk buildings. A deflector plate at the sprinkler head achieves the desired coverage (figure 6.19).

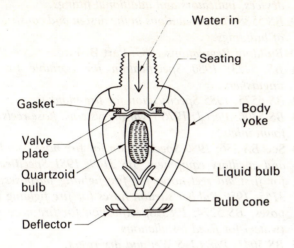

Figure 6.19   A quartzoid sprinkler head

Sprinkler heads commonly operate on the breaking of a soldered link as a result of heat action or by the fracture of a bulb of high expansion liquid as shown in figure 6.9.

Permanent hoses are fed water by relatively large diameter pipes, and water release is by manual opening of an adjacent stop valve. Hoses are generally 25 mm bore diameter while the supply pipe is typically 64 mm minimum bore. This type of fire fighting equipment tends to be used to protect the integrity of escape routes in assembly, institutional or other buildings of high occupancy.

Risers are used to carry water to the various levels in multi-storey buildings for the connection of fire brigade hoses. In the same way as the sprinkler installation, risers are either wet or dry to allow for seasonal variations in temperature.

The principles of the two arrangements are illustrated in figure 6.20 which shows the use of the fire engine as a pump in the dry system. With the dry system, a special valve releases the air displaced in the piping as water enters the riser, this valve providing a seal to water when all of the air has been expelled. In contrast, wet risers provide an immediate water supply to the hose connection

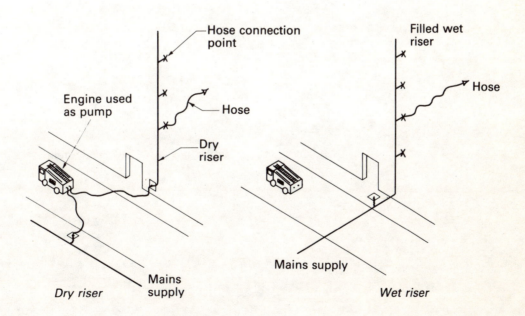

Figure 6.20   Dry and wet riser systems

by the manual opening of a control valve. In tall buildings pressure assistance may be necessary to ensure a satisfactory delivery pressure at the highest point, and equipment similar in nature to that used for the assisted delivery of cold water as described in chapter 1 is applied. In both the wet and dry arrangements, a minimum 100 m diameter riser is used.

REFERENCES

1 Chartered Institution of Building Services (CIBS), *IHVE Guide B2, Air Conditioning Requirements.*

2 Chartered Institution of Building Services (CIBS), *IHVE Guide A10, Condensation and Moisture Problems.*

3 Chartered Institution of Building Services (CIBS), *IHVE Guide B3, Air Conditioning Systems and Equipment.*

4 Chartered Institution of Building Services (CIBS), *IHVE Guide B14, Refrigeration and Heat Rejection.*

5 *BS 5655: Part 6: 1985 Code of practice for the selection and installation of lifts.*

6 *BS 5655: Part 1: 1979 Safety rules for the construction and installation of electric lifts.*

7 *BS 5655: Part 2: 1983 Specification for hydraulic lifts.*

8 *BS 5655: Part 7: 1983 Specification for manual control devices, indicators and additional fittings.*

9 *BS 5588 Fire precautions in the design and construction of buildings.*

10 *Building Regulations 1985.* Part B, Fire.

11 *BS 5423: 1980 Specification for portable fire extinguishers.*

12 *BS 6575: 1985 Specification for fire blankets.*

13 *BS 5306: Part 1: 1976 Hydrant systems, hose reels and foam inlets.*

14 See *BS 336: 1980 Specification for fire hose couplings and ancillary equipment; BS 3169: 1981 Specification for first aid reel hoses for fire fighting purposes; BS 3165: 1959 Rubber suction hoses for fire fighting purposes; BS 5274: 1985 Specification for fire hose reels (water) for fixed installations.*

15 *BS 5041: Parts 1–5 Wet and dry risers.*

7 ELECTRICAL INSTALLATIONS

INTRODUCTION

Electricity supplies for distribution through the National Grid are high voltage supplies: 400,000, 275,000 or 132,000 volts (400, 275 or 132 kV). These voltages are reduced by stepdown transformers to 11,000 and 6000 volts at sub-stations, but even at this reduced rate the supplies are too large for most buildings. Local transformers bring further reductions to 415/240 volt supplies which have three phases (or live conductors) and a neutral.

Figure 7.1 shows that the voltage potential between the live conductors is 415 volts and between the live and neutral 240 volts.

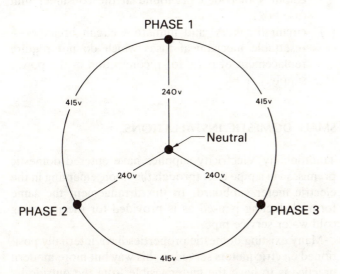

Figure 7.1 Voltages — three-phase four-wire supplies

A single-phase supply is most common for residential use (one live and one neutral — 240 volts potential between), while industrial and larger consumers may take the three-phase supply.

The bulk of the electricity generated is alternating current (AC); this means that the current flow direction is constantly reversing at the rate of 50 times per second. On larger installations, circuit breakers provide the function of a fuse but have the advantage they they can be reset without the replacement of any parts. With three-phase supplies a fuse or circuit breaker will be fitted to each phase.

Large consumers tend to be provided with the three-phase supply from which single-phase branch connection can be made. Figure 7.2 shows a typical busbar system which may be used to carry the three-phase supply. This consists of four copper rods threaded through rigid insulators to space the conductors apart, and a covering of trunking to protect against electrocution.

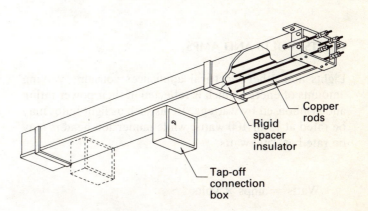

Figure 7.2 A busbar conductor

Figure 7.3 shows an alternative to the busbar detail where insulated PVC cable is used instead of the bare copper rods.

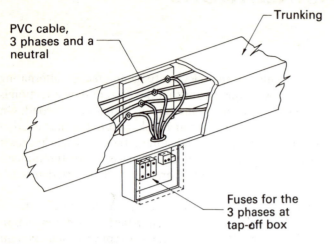

PVC cable,
3 phases and a
neutral

Trunking

Fuses for the
3 phases at
tap-off box

Figure 7.3   Encased PVC conductors

Within the design of any electrical installation the regulations of the Institution of Electrical Engineers (IEE)[1] provide a basis for quality standards which are used throughout the building industry.

WATTS, VOLTS AND AMPS

Lights and other electrical appliances consume varying amounts of energy which is reflected in their power rating and symbolised by wattage. Items such as light bulbs may be rated at 25 or 100 watts, while immersion heaters may be rated at 2500 watts.

$$\text{Watts} = \text{amps} \times \text{volts}$$

and therefore

$$\text{amps} = \frac{\text{watts}}{\text{volts}}$$

When 1 watt is consumed, 1 amp of current will flow across a 1 volt supply.

If a 2 kW immersion heater is used on a 240 volt supply, the current taken in amps is

$$\frac{\text{watts}}{\text{volts}} = \frac{2000}{24} = 8.33 \text{ amps}$$

The current in amps will determine the size of cable taken to a point of electricity consumption. Cable sizes[2] are reflected in cross-sectional area in square millimetres. For example, a 2 amp current could be conveyed by a 0.5 mm^2 area conductor, while a 20 amp current would require a conductor of 2.5 mm^2 area.

Fuses of different rating will be used on the live wire, the size of fuse being related to the current supplied. Fuses contain wire of smaller diameter than the conductor on which they are placed, creating a weak link in the circuit. In the event of an abnormality in the current, it is the weak link which will fail by melting, causing the power supply to cease.

The fuse may take various forms:

(a)   cartridge — rating marked and coloured (as commonly found in the household plug);
(b)   porcelain rewirable — rating marked and sometimes colour spot coded (as found at the consumer unit fuse box);
(c)   circuit breakers[3] and miniature circuit breakers — resettable mechanical fuses which do not require replacement of parts for reconnection of the power supply.

SMALL DOMESTIC INSTALLATIONS

Traditionally, electricity supplies have entered domestic premises via a pipe duct through the floor, emerging in the electric meter cupboard. In this arrangement the same form of ducting is used as is provided for the incoming cold water service pipe.

Many existing domestic properties have internally positioned electric meters served in this way but more modern practice is to have the meter visible from the outside by fitting inspection boxes to the outside wall of the house. This allows the meter to be read without disturbance of the occupants and also ensures access for the Electricity Board meter reader. Irrespective of the meter position, there are a number of other component parts of the incoming supply controls which are located at the consumer board. Figure 7.4 shows a typical installation consumer board.

The wrapped incoming Electricity Board supply is provided with a 100 amp fuse before cable tails are taken to the electric meter and further cable tails connect the supply to the consumer control and fuse box. Indication of

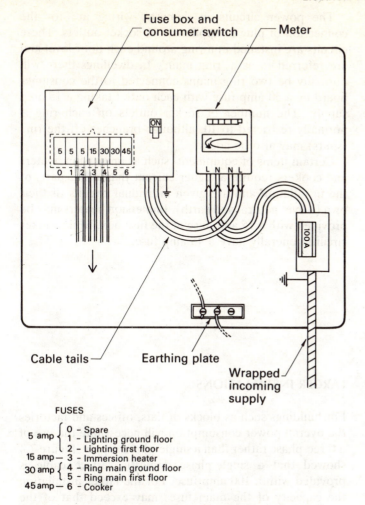

FUSES

5 amp	0 – Spare 1 – Lighting ground floor 2 – Lighting first floor	
15 amp	3 – Immersion heater	
30 amp	4 – Ring main ground floor 5 – Ring main first floor	
45 amp	6 – Cooker	

Figure 7.4   Electricity entry into the building — the domestic consumer board

the likely use of the fuses is given in the illustration.

As well as having a purpose-made earth via the earthing plate, a further earth is often taken on to the metal gas supply pipework where gas is supplied to the premises.

Distribution into the dwelling is generally through three types of circuit: lighting, power and equipment power. Lighting is a 5 amp circuit with a live and a neutral taken to each ceiling rose (figure 7.5).

Electrical cabling consists of a conductor (usually copper) surrounded by various types of insulation: PVC, plastic, rubber or mineral compounds.

Classification of cable is generally by the number of conductors within the cable — for example, single, twin, twin and earth, triple and earth — and by the cross-

sectional area of the conductor: 1.0, 1.5, 2.5, 4.0, 6.0, 10.0, 16.0 or 25 mm^2. The different diameters of cable are to suit the different current ratings to be carried. Mineral insulated copper covered cables (MICC) are used particularly where resistance to high environmental temperatures or protection from possible fires may be necessary. Single core cables are for use specifically in conduit which will offer some protection to the cables and prevent damage. The most widely used conduit is screwed steel, although PVC and copper may also be used. Metal conduits can also provide the useful function of an earth to the system.

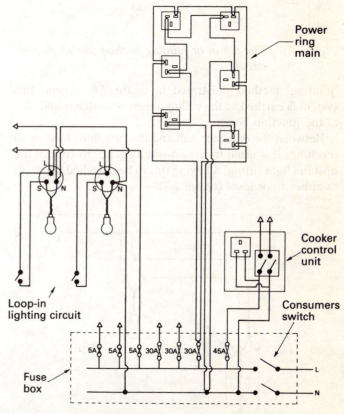

Figure 7.5   A diagrammatic representation of the electrical installation of a dwelling

Cabling to the lighting circuit is generally 1.0 mm^2 cross-sectional area. The illustration shows the switch drop(s) from the live wire but does not show the earth connection which is normally made to all ceiling roses and all switches.

Figure 7.5 shows the 'loop-in' arrangement of forming a lighting circuit which may be compared with the 'tap-in' or

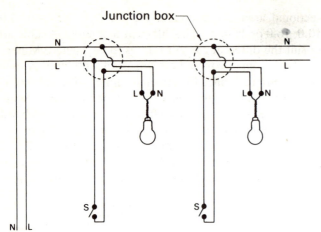

Figure 7.6   The tap-in or jointing method for a lighting circuit

'jointing' method illustrated in figure 7.6. Again, this system is earthed at the ceiling roses, at switches and also at the junction boxes.

Between the entrance hall and the first floor landing of dwellings it is usual to fit two-way switches to control the upstairs light fitting, allowing this to be switched on or off at either floor level (figure 7.7).

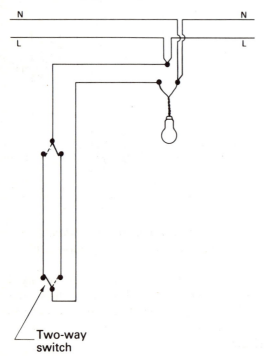

Figure 7.7   Two-way switching for lights

The power circuits in domestic wiring are for the connection of plug-in appliances at socket outlets. These circuits are installed in a ring around each floor level and are referred to as a 'ring main'. In dwellings there will normally be two ring mains connected to the consumer board by a 30 amp fuse with each outlet taking a 13 amp supply. The number of socket outlets on each ring is normally restricted to 10, although branches off the ring (spurs) may also be added.

Certain items of equipment[4] such as immersion heaters and cookers require a higher current than other parts of the installation and are given individual circuits of their own (a live, neutral and earth). Immersion heaters may be provided with a 15 amp fuse at the fuse box, while cooker circuits generally have a 45 amp fuse.

LARGER INSTALLATIONS

For buildings such as blocks of flats, offices and factories the overall power consumption will necessitate the use of a three-phase rather than a single-phase supply. Figure 7.4 showed that a single-phase domestic service cable is provided with a 100 amp fuse. For the larger installation the capacity of the main fuses may exceed that of the domestic installation by tenfold, probably employing a circuit breaker to provide this function.

There may be many user outlets to serve, and particularly in multi-storey structures there will be items in common use which will require a power supply: lifts, the heating boilers and perhaps air conditioning.

In these types of installation a three-phase busbar panel[5] can present a means of easy connection for single-phase rises to the various levels within the building. A final sub-circuit will extend into the user's premises via a consumer control board where the isolation switch and fuses are located. Single-phase risers for a three-phase busbar panel are shown in figure 7.8.

An alternative to this would be to take a three-phase supply busbar throughout the full height of the building and tap-off single-phase supplies as required to each floor level (figure 7.9).

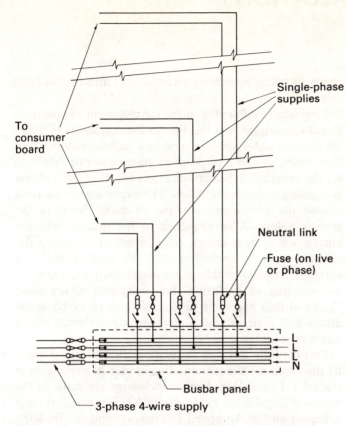

Figure 7.8   Multi-storey single-phase risers from a three-phase service mains

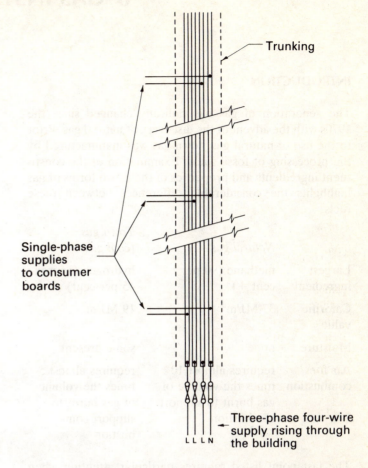

Figure 7.9   A multi-storey rising three-phase supply with single-phase branch connections

REFERENCES

1 The Institution of Electrical Engineers, *Regulations for the Electrical Equipment of Buildings.*
2 *BS 4746: 1984 PVC insulation and the sheath of electric cables.*
3 *BS 842: 1965 (1980) AC voltage operated earth leakage circuit breakers.*

4 *BS 3456 Specification for safety of household and similar electrical appliances.*
5 *BS 5486: Parts 1, 2, 11, 12 and 13 Specification for factory built assemblies of switchgear and control gear for voltages up to and including 1000 v ac and 1200 v dc.*

8 GAS INSTALLATIONS

INTRODUCTION

The generation of gas has radically changed since the 1970s with the advent of the discovery of natural gas. Prior to the use of natural gas, town gas was manufactured by the processing of fossil fuels. Examination of the constituent ingredients and properties of these two forms of gas highlights the considerable differences between these fuels.

	Natural gas	Town gas (coal gas)
Largest ingredient	methane (90 per cent +)	hydrogen (approx. 45 per cent)
Calorific value	37 MJ/m^3	19 MJ/m^3
Moisture	none	some present
Air for combustion	requires almost 10 times the volume of gas burnt to support combustion	requires almost 5 times the volume of gas burnt to support combustion

The last point listed requires particular attention when gas-burning appliances are introduced to existing property for the first time.

Older premises that undergo refurbishment schemes may have to be provided with additional air vents to ensure that an adequate supply of air is provided, allowing complete combustion of the fuel. The use of double glazing, secondary glazing and modern well-fitting windows highlights the need to create new positions of air entry as a result of the much reduced natural air penetration.

Safety is of paramount importance to the user of gas appliances and extensive controls have emerged, many of which have been incorporated into the *Building Regulations*.

Particularly influential in the development of constructional regulations was the Ronan Point disaster of 1968. This concerned a gas explosion in a multi-storey block of flats which caused a disproportionate amount of damage for the severity of the explosion as a result of the building possessing design weaknesses. The explosion at one level caused the progressive collapse of many floors in the storeys below. A consequence of the disaster was the emergence of new design rules aimed at localising the damage from an explosion, and the undertaking of a survey of similar buildings to assess their capability of withstanding an explosion. Following this survey some blocks of flats had their gas installations removed, while others had their structural members strengthened where this was considered viable.

Part D of the *Building Regulations 1976* included specific provision for the structural resistance of explosions in clauses D17 and D18. Similar inclusions are made in the *Building Regulations 1985*, clause A3[1] (Disproportionate collapse) and in Approved Document A to the Building Regulations.

Clause A3 indicates that the rules listed relating to disproportionate collapse apply to buildings of 5 or more storeys (including basements), and to public buildings which have clear spans in excess of 9 metres. Localisation of the explosion is said to have been achieved if the damage caused is restricted to the floor of the explosion and to the floors immediately above and below, and also if the damage on those storeys is restricted to 70 m^2 or 15 per cent of the area of the storey (whichever is the less).

In addition to the *Building Regulations*, the Gas Industry has its own set of safety regulations as listed in the *Gas Safety Regulations 1972*. The scope of these regulations extends to the installation of piping, meters and appliances.

CONNECTION OF THE GAS SUPPLY

A number of means may be employed to provide a gas service pipe connection to a low rise dwelling, including:

(a)   service entry below ground;
(b)   service entry above ground;
(c)   service entry below floor;
(d)   service entry via a meter box; and
(e)   service entry into a garage (no more than 2 m of pipe to be exposed before connecting to the meter).

Figures 8.1 and 8.2 illustrate some of these possibilities.

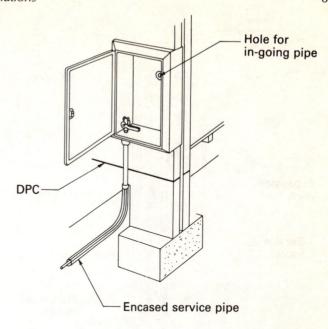

Figure 8.2   Gas service pipe entry via a meter box

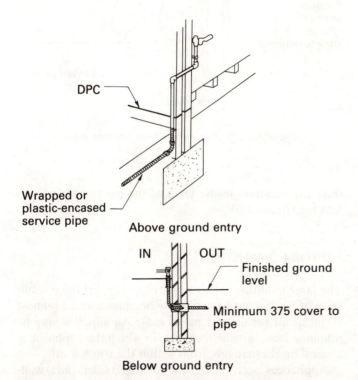

Figure 8.1   Gas service pipe entry above and below ground

Where the incoming service pipe from the mains was to convey town gas, a shallow fall (1:120) was provided to the service pipe to direct condensate back towards the main.[2] Some town gas service pipes were also provided with a condensate collector to allow removal of condensate which accumulates in the pipe. However since natural gas does not contain the moisture of town gas, service connections no longer have to be laid to falls. Irrespective of whether town or natural gas is used, a minimum cover of 375 mm to finished ground level is still observed.

If the gas supply is to be taken through to the different levels of a multi-storey building, for example a block of flats, a protected and ventilated shaft may be used (figure 8.3).

The gas meter is provided with a pressure governor to stabilise and regulate the delivery pressure; a typical meter is shown in figure 8.4.

Supply pipes are generally in wrapped steel or plastic-encased steel. Where the service or distribution pipework passes through structural elements such as walls and floors, a pipe sleeve is used to act as a duct for the pipe. Sleeves are sealed around the pipe with an appropriate material which may need to be of firestopping quality where compartment walls, compartment floors or protected shafts are penetrated.

To check that the correct amount of gas is being delivered to the various appliances, a pressure test is carried out using a simple U-tube water gauge (figure 8.5). The water gauge is also used to test for leakage. Where a pipe length is to be tested for leakage, air can be pumped into the pipe using simple hand bellows until a prescribed water gauge pressure is registered. Following

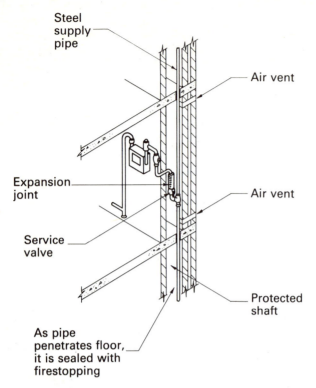

Figure 8.3   Gas supplies to multi-storey buildings

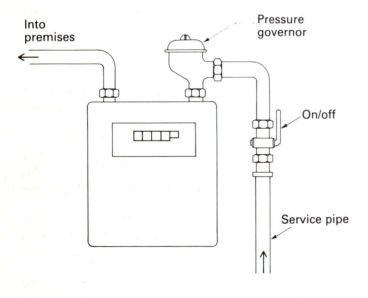

Figure 8.4   The gas meter with pressure governor

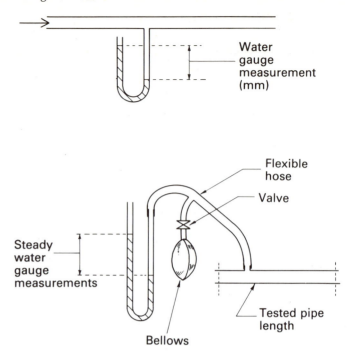

Figure 8.5   Water gauge tests for gas piping

this, any variation in the level of the gauge will indicate leakage (figure 8.5).

AIR FOR COMBUSTION[3]

The large quantities of air required to support the combustion of natural gas have already been mentioned (almost 10 m^3 of air for every 1 m^3 of gas). Air supplies may be obtained from outside the room in which the appliance is housed or alternatively from within the room itself.

Appliances such as central heating boilers and wall-mounted space heaters may obtain their air for combustion from outside the building and are consequently termed 'room-sealed'. A balanced flue arrangement allows an appliance to be called room-sealed since the air for combustion and the combustion wastes are fed through a special flue mounted in the external air. Figure 8.6 illustrates the principle of the balanced flue detail.

Other gas appliances obtain their combustion air from within the room in which they are located, either directly from the room itself or indirectly from adjacent connected

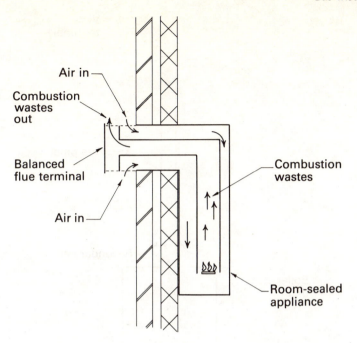

Figure 8.6   A balanced flue detail to a room-sealed appliance

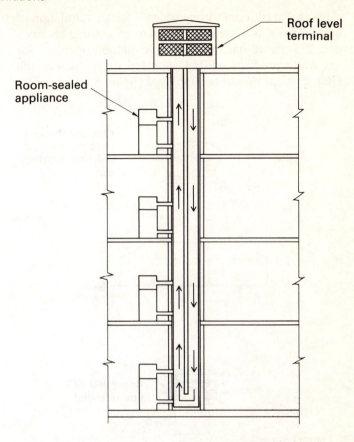

Figure 8.7   The 'U-duct' principle

spaces. As well as the specific design ventilation areas through which external air may be drawn in to support combustion, there will generally be 'advantitious ventilation' provided through windows and around doors. Research by British Gas[4] has shown that despite attempts to control the room air change rate by weatherstripping and double glazing, it is virtually impossible to reduce the advantitious ventilation openings of habitable rooms to less than 3500 mm^2.

In multi-storey construction, room-sealed appliances may be provided with a 'U-duct' system in the provision of combustion air and for the expulsion of combustion wastes (figure 8.7).

THE EXPULSION OF COMBUSTION WASTES[5,6]

There are a number of ways in which the expulsion of combustion products from gas appliances may be discharged. The uses of a balanced flue and the U-duct principle have already had mention, but other means include:

(a)   discharge via a flue chimney designed for the expul-

sion of the combustion wastes from fossil fuels;
(b)   discharge via flue pipes;
(c)   discharge via a gas flue block system;
(d)   discharge via communal flue ducts.

In respect of the use of chimneys which were originally designed to take the wastes from the burning of fossil fuels, two situations may occur:

(i)   the chimney flue is an unlined brickwork detail and necessitates the insertion of a flexible flue lining (such as Kopex) fitted with an approved vent terminal; or
(ii)   the flue is a more modern lined type which requires no specific treatment.

When unlined chimney flues are provided with a flue liner or converted to the expulsion of gas products, the wrong type of flue terminal will sometimes be fitted. The clayware 'mushroom' fitting which is for air ventilation of unused flues, is sometimes added to an existing chimney pot. This detail has insufficient ventilation openings for gas waste products and an approved detail such as the GCI terminal should be employed (figure 8.8).

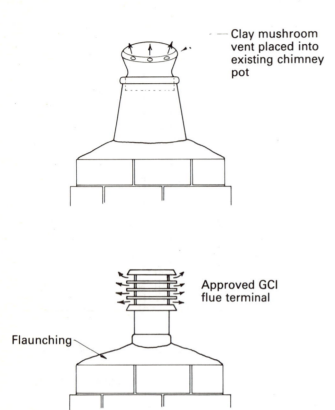

Figure 8.8   Discharges from open fire flues — terminal details

Circular flue pipes may be in a variety of materials,[7] including sheet metal (BS 717: 1970), stainless steel (BS 1449: 1983), asbestos–cement (BS 567: 1973), cast iron (BS 41: 1973) and vitreous enamelled steel (BS 1344: 1965 and 1967).

The Building Regulations 1985 make specific demands on the details employed where flue pipes pass through elements of the building fabric (Approved Document J, J1/2/3, Section 1.32), and stipulate the minimum diameters of flue pipes used in open flue arrangements. Figure 8.9 shows a diagrammatic representation of a typical open flue detail, and indicates the point of secondary air entry for assisting in the expulsion of the combustion wastes.

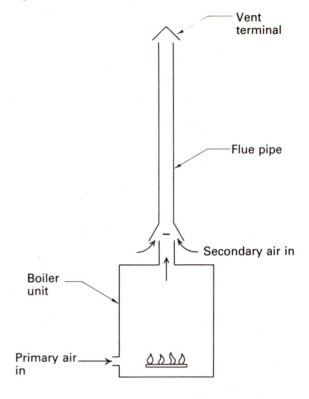

Figure 8.9   Pipe open flues for gas appliances

In residential construction, provision may be made for a flue to serve a lounge gas fire by use of a gas flue block system. This type of flue consists of blocks of similar dimensions to the blockwork used for the inner skin of the cavity wall. They are therefore easily bonded into the inner skin and eliminate the chimney breast projection. A number of different profiles of block are used in the flue as shown in figure 8.10:

recessed — to take the back of the gas fire
taper — to act as the throat entry to the flue
straight blocks
rakers — to assist the draw of the combustion wastes
connector — to allow the connection of a circular asbestos–cement pipe which carries the combustion wastes from the flue blocks to the roof ridge terminal

It should be noted that, because of the limited volumetric capacity of this flue, it is capable only of transmit-

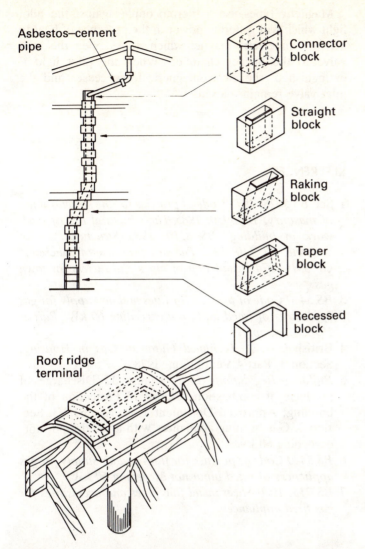

Asbestos–cement pipe

Connector block

Straight block

Raking block

Taper block

Recessed block

Roof ridge terminal

Figure 8.10   A typical gas flue block system (courtesy of Marley Flue Systems)

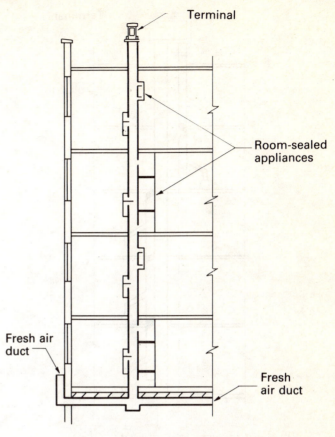

Terminal

Room-sealed appliances

Fresh air duct

Fresh air duct

Figure 8.11   The 'SE-duct' flue

ting the discharges from gas appliances and is not capable of taking the discharges from an open fire (coal burning).

In multi-storey buildings where gas appliances are to be used on successive floors, various types of communal system may be used to carry the combustion wastes. These arrangements are sometimes also used to supply the primary air for combustion, such as in the 'SE-duct' flue (figure 8.11). Where appliances are not room-sealed, the 'shunt-duct' principle may be employed, as shown in figure 8.12.

FLAME FAILURE DEVICES

Safety in gas usage is a primary concern. A means of controlling the flow of gas in the event of an appliance pilot light being extinguished is one important in-built safety feature. All automatic appliances including central heating boilers and instantaneous gas water heaters must have a flame failure device. These may be:

(a)   bi-metal types; or
(b)   magnetic types.

In the bi-metal detail, a metal strip consisting of steel alloy and brass is positioned to touch the flame of the pilot light. While the pilot light burns the relative expansion properties of the two metals cause distortion and this

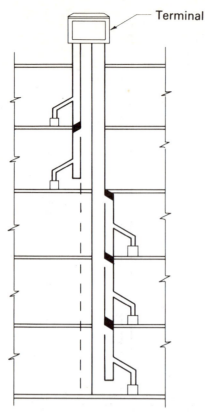

Figure 8.12   The shunt-duct flue arrangement

opens a valve, allowing gas into the burner chamber. Should the pilot light be extinguished, the contraction of the bi-metal strip causes the gas inlet to be closed.

Magnetic types use a thermocouple against the pilot light which will generate power in the presence of heat to activate an electro-magnet which draws open the inlet valve to the burner chamber. When the pilot light is extinguished, the electromagnetic force ceases and the inlet valve remains closed.

REFERENCES

1 See *BS 5628: 1978 Code of practice for the structural use of masonry; BS 5950: 1985 The structural use of steel-work in building; BS 8110: 1985 Structural use of concrete; CP 110: 1972 The structural use of concrete.*
2 *CP 331: 1974 Installation of pipes and meters for town gas.*
3 *BS 5440 Code of practice for flues and air supply for gas appliances of rated input not exceeding 60 kW. Part 2: 1976 Air supply.*
4 British Gas, *A Technical Guide to Gas in Housing*, Section 3, Part 4 Ventilation, 1978.
5 *Building Regulations 1985*, Clause J2 Discharge of products of combustion; Clause J3 Protection of the building; Approved Document J, clauses J/1/2/3, Section 2, Gas burning appliances with a rated input not exceeding 60 kW.
6 *BS 5440 Code of practice for flues and air supply for gas appliances of rated input not exceeding 60 kW.*
7 *BS 715: 1970 Sheet metal flue pipes and accessories for gas fired appliances.*

9 REFUSE DISPOSAL

INTRODUCTION

A number of factors are influential in relation to the collection of refuse:

(a) the nature of the waste — perishable or non-perishable
(b) the nature of the premises generating the waste
    — domestic
    — industrial
    — commercial
(c) the location of the point of generation
    — height from the ground
    — position on the floor layout

The nature of waste generated by dwellings (a primary source of refuse) has changed in the last two decades in that waste is becoming less dense but of greater bulk. There has also been a tendency towards the use of non-reusable packaging, particularly following the advent of plastics.

This chapter concentrates in the main on the problem of collecting waste from residential buildings, since the nature of waste produced by industrial and commercial buildings is often of a specialised nature.

The geometry of the premises served by refuse collection will have a considerable influence on the design of the collection method. Height is particularly influential.

Clause H4 of the *Building Regulations 1985* outlines the need for refuse chutes to serve buildings over four storeys in height, as collection of individual volumes of refuse could not be efficiently served by manual collection from higher levels.

Code of Practice 306 previously helped to define the types of container which could be used to store refuse:

(a) individual refuse containers (such as dustbins);

(b) refuse storage containers (movable, capacity not exceeding 1 m^3);
(c) bulk containers (capacity over 1 m^3 but less than 9.17 m^3).

In respect of the use of dustbins, an interesting observation is that the *Building Regulations 1985* refer to the provision of access to a dwelling for the removal of refuse from containers of at least 0.12 m^3 capacity. Individual refuse containers to the definitions of CP 306 are of capacity not exceeding 0.11 m^3, and this suggests that at least two dustbins per household would be necessary. However BS 5906: 1980[1] has subsequently revised the titles and the capacities of refuse containers, replacing CP 306 as the current design reference:

(a) individual waste containers (not exceeding 0.12 m^3 capacity);
(b) communal waste storage containers (0.75–1 m^3 capacity); and
(c) bulk waste containers (1–30 m^3 capacity).

SOLID WASTE DISPOSAL SYSTEMS

A chart may be used to summarise the methods available for the collection and disposal of solid waste from residential premises (figure 9.1).

Storage and Direct Collection

The depositing of refuse manually by residents into various sizes of container has already been mentioned as a satisfactory disposal method for buildings of no more than four storeys. For taller buildings, however, a chute may represent an efficient method of disposal.[2] A typical refuse chute is illustrated in figure 9.2, which shows the main features of the chute design. The chute needs to

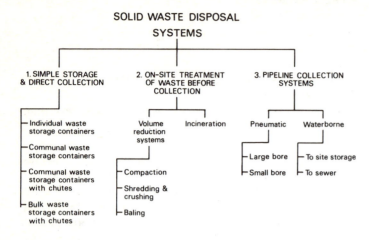

Figure 9.1   Solid waste disposal systems

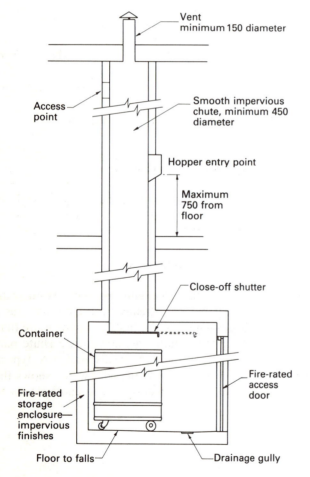

Figure 9.2   The component parts of a refuse chute

present a smooth, impervious and non-combustible finish to the refuse conveyed. Suitable materials for chutes include:

(a)   clay pipes to BS 65;
(b)   concrete pipes to BS 556;
(c)   steel sheet to BS 1449 Part 1, and galvanised steel to BS 729;
(d)   stainless steel sheet to BS 1449 Part 2.

On-site Treatment of Waste

With installations of refuse collection which collect large volumes of refuse, it may be considered viable to install a means of volume reduction. Volume reduction may be by compaction, shredding, baling or incineration.

Automatic rotating turntables on which sacks or other containers may be placed provide a means by which the period between attendance for collection may be reduced. Such turntable systems also often incorporate a ram compactor which can achieve volume reductions of the order of 4:1, thus increasing the time period further between attendance for refuse removal. Incineration is even more effective in achieving volume reductions, with ratios of 10:1 being typical. Incineration does however mean high installation costs for machinery, high operation costs in terms of energy usage and the need to comply with legislation such as the Clean Air Acts.[3]

Pipeline Collection Systems

It should be appreciated that many of the pipeline systems have a limited capacity to deal with, in terms of the percentage of the total quantity of refuse generated. Some pipeline systems may only be capable of dealing with perishable waste and this may be less than a quarter of total refuse.

However there are some systems which convey all of the generated refuse and one example of such a system is the pneumatic pipeline. This is a refuse duct of variable bore which collects the refuse from the foot of a series of refuse chutes under the action of air suction. The pressurised ducts radiate from a central collection point and this system has been used to serve groups of multi-storey buildings. The suction piping does not normally exceed 500 mm bore and cast iron is the most widely used material. On arrival at the collection point the refuse is incinerated or ram-compacted.

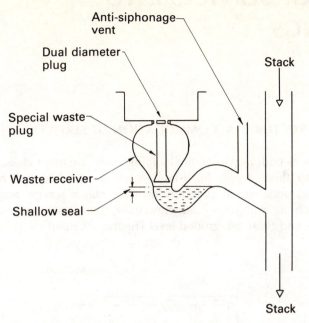

Figure 9.3   The sink receiver of the Garchey system

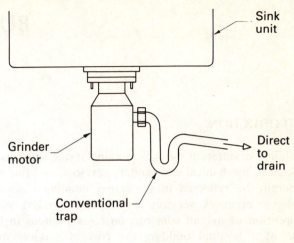

Figure 9.4   A sink grinder

Wet transfer of refuse may be a less expensive method of pipeline collection but this method cannot deal with the total amount of refuse generated. Two examples of the wet pipeline technique are the Garchey system and the sink grinder system.

The Garchey system can deal with around 40–50 per cent of total refuse, while the sink grinder handles basically perishable material amounting to approximately 10–15 per cent of total refuse. The Garchey method has had very little application in the UK but is more widely used in other European countries. It is known as a waterborne-to-site storage technique since the material collected is held in a storage chamber for further processing or collection. Figure 9.3 shows the bulbous receiver of the system which is located at the kitchen sink. The illustration shows the need for anti-siphonage pipework to protect the shallow trap from seal loss.

This system is able to accommodate tins and small bottles of maximum dimension 75 mm diameter and 150 mm in length. Operation of the outlet from the receiver releases the mixture of solids and liquids which are conveyed by a vertical stack to the holding chamber. From this chamber the material is either incinerated on site or collected by tanker.

Sink grinders,[4] in contrast, deal largely with vegetable preparation waste by pulverising it into a pulp which, when mixed with tapwater, is readily carried by a conventional waste pipe to the drain. Because of the pulp nature of the discharge from the grinder, bottle traps are not suitable for the outlet and the waste does not discharge into a gully but directly into the drain (figure 9.4).

REFERENCES

1 *BS 5906: 1980 Code of practice for the storage and on-site treatment of solid waste from buildings.*
2 *BS 1703: 1977 Specification for refuse chutes and hoppers.*
3 Clean Air Acts.
4 *BS 3456: 1979 Food waste disposal units*, Section 3.8.

10 THE INTEGRATION OF SERVICES INTO BUILDINGS

INTRODUCTION

Buildings of different use will contain varying amounts of electrical, mechanical and sanitary services, and this will generally be reflected in the initial building costs. In domestic premises, services may represent only a small proportion of overall construction costs whereas in the case of a hospital building the cost of services may outweigh the costs of the building fabric.

In service-intensive buildings, the architect and services engineer should collaborate in the development of the design in an attempt to achieve:

  (i) easy access for service installation, providing the minimum disruption to other building activities;
 (ii) easy access for maintenance and repair, providing the minimum disruption to the building occupant; and
(iii) isolation of services for the safety of occupants and to minimise nuisance in operation.

There are many different ways in which pipe, cable and box-ducted services may be routed around the building.[1] To aid the classification of possibilities, a division may be made into structural and non-structural methods of accommodation.

Structural accommodation will be particularly influential in design, as this classification represents ducts and recesses built into the fabric of the building. Non-structural accommodation includes methods of containment attached to the structural elements of the building such as suspended ceilings, raised floors and surface trunking.

STRUCTURAL ACCOMMODATION OF SERVICES

Ducts built as part of the building fabric are often classed into three types: horizontal, vertical and lateral. The horizontal group generally includes major service ways such as walkways and crawlways, these occurring at ground or at sub-ground level (figures 10.1 and 10.2).

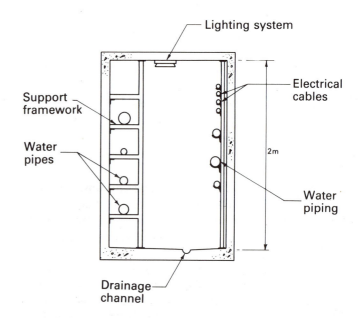

Figure 10.1   The walkway duct

These horizontal ducts contain the bulk of the pipe and cable services which are routed around the floor plan of the building, and from these vertical ducts will transfer the services to each floor level.

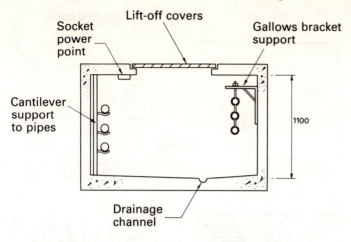

Figure 10.2   The crawlway duct

Within the duct there will be a logical arrangement of the items conveyed. Hot pipes, for example, will be kept away from the incoming cold water supplies, and water-carrying pipes will not be located above the electrical cabling.

Walkways have their own lighting system and electrical power points at intervals for the attachment of power tools. Illumination of crawlways will be by plug-in portable lighting and access will have to be provided at regular intervals (if not continuously) because of the greater restrictions on movement.

Vertical ducts connecting the various floor levels may be grouped into two types: internal access vertical ducts and external access vertical ducts (figure 10.3).

Internal access vertical ducts are sufficiently large to allow an operative to stand inside while attending the services. External access vertical ducts, in contrast,

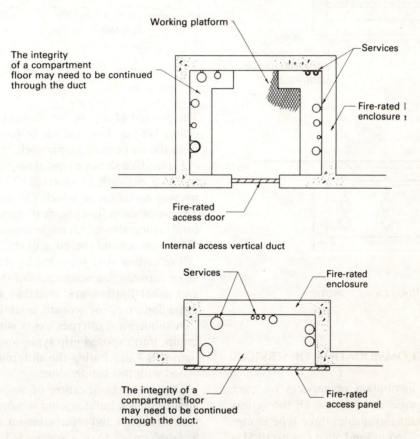

Figure 10.3   Internal and external access vertical ducts

normally have a full-width removable access cover, as the operative will stand outside the duct while attending the services.

On reaching each floor level, the conveyance may continue in structural accommodation in the form of floor trenches (figure 10.4). These are lateral means of distribution.

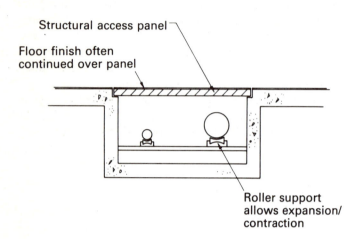

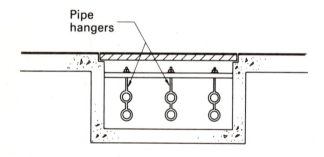

Figure 10.4   Floor trenches

NON-STRUCTURAL ACCOMMODATION OF SERVICES

Non-structural lateral distribution of services on each floor may often be via suspended ceilings. Of the variety of types of suspended ceiling available, three type groups emerge, namely jointless, jointed and open (figure 10.5).

Jointless types are not usually employed for the conveyance of services since the continuous monolithic ceiling finish does not readily permit access to the ceiling void.

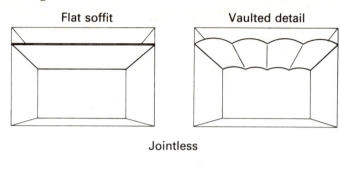

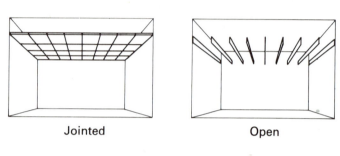

Figure 10.5   Types of suspended ceiling

The jointed types, in contrast, allow easy access to the ceiling void as these use tile or panels which are removed from the suspension framework.

Figure 10.6 shows a typical suspension assembly consisting of a network of inverted 'T' sections in light-gauge pressed metal on to which the ceiling tile finish is laid. This suspension framework is often shotfired to the structural ceiling above. An angle section replaces the inverted 'T' profile around the edge of the room.

The ceiling void provided by this type of access has a large capacity for accommodation.[2] Many attractive tile and panel finishes are available with many of the tiles providing excellent acoustic qualities.

A subdivision of types exists within the jointed ceiling group, into exposed grip types and concealed grid types, figure 10.7 illustrating the differences in tile fixing associated with this subdivision.

The open classification of suspended ceiling may be applied in new buildings but is more commonly associated with alteration and refurbishment schemes. In the system a 'false' rather than a suspended ceiling may provide a truer description of the detail. Slats, boards or grids are used to form the ceiling layer, leaving the ceiling void open to view. However, as these details are brightly

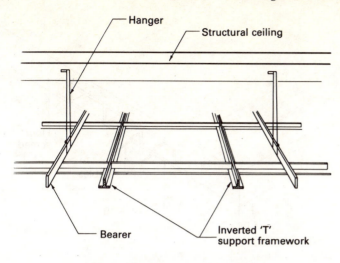

Figure 10.6 A typical support suspension framework for a suspended ceiling

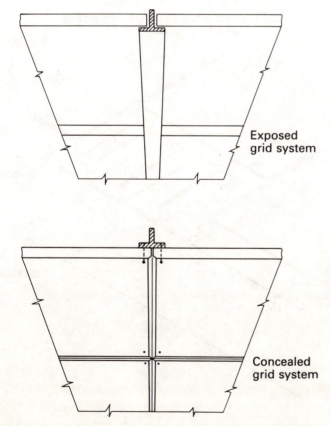

Figure 10.7 Exposed grid and concealed grid types of jointed suspended ceiling

coloured while the void walls and true ceiling are darkly finished, the eye is drawn to the brighter colours and the void should be little noticed. This system often provides a means by which the floor-to-ceiling dimension is reduced to improve aesthetics.

If suspended ceilings are to be used in multi-storey construction as the main route for service distribution on each floor, the effect of the ceiling void on the overall height of the building should be appreciated.

An alternative to the use of suspended ceilings may also affect the overall height of building in a similar way — the raised floor detail.

Raised floors were originally developed to house the extensive cabling associated with computer installations. Deep cavity raised floors have a floor cavity in the region of 250 mm created by the use of jack posts supporting structural floor panels.

As computer installations require a controlled atmospheric environment free from dust, the edges of the floor panels have to be provided with sealant material and the integrity of the seal is preserved around the perimeter of the room, usually by the use of flexible PVC skirtings dressed over the panels. Computers also exert considerable dead weight on the floor, necessitating a panel strength capacity of around 500 kg/m^2. Reinforcing beams are sometimes used to span the tops of the jack posts, thereby allowing larger floor panels to be used.

Figure 10.8 shows a jack post detail which supports four corners of floor panels across the general area of a raised floor.

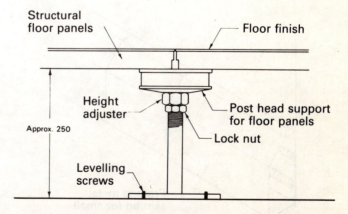

Figure 10.8 A jack post for a deep cavity raised floor system (courtesy of Robertson Raised Floor Systems)

An alternative to the deep cavity raised floor is the shallow cavity solution. Such systems may be fabricated *in situ* or be purpose-made (figure 10.9).

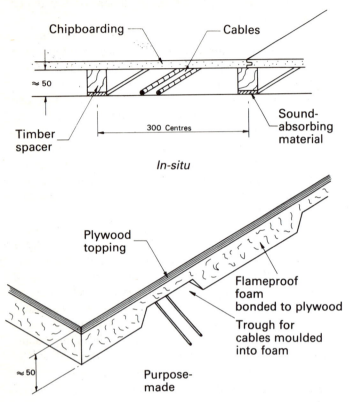

In-situ

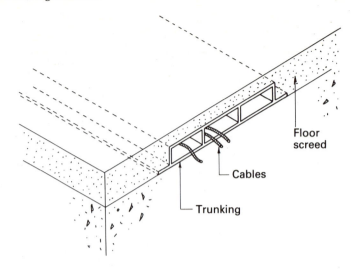

Figure 10.9 In-situ *and purpose-made shallow cavity raised floors*

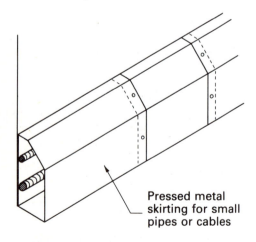

Figure 10.10   Skirting trunking in pressed metal

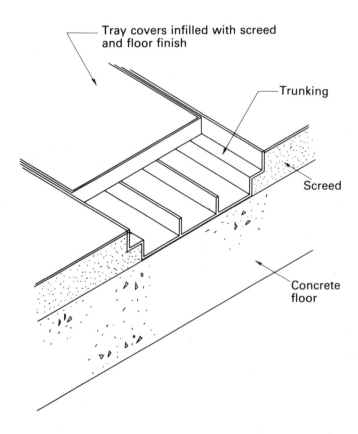

Figure 10.11   Trunking embedded in the floor screed

Both types of shallow cavity detail are restricted to the accommodation of cable services but may be used in commercial premises where extensive desk top machinery or the like is to be used.

If a shallow cavity raised floor is not to be used, skirting trunking or embedded trunking may be used (figures 10.10 and 10.11).

FIRE PRECAUTIONS AND SERVICE DISTRIBUTION

Service ducts and other service ways which are routed around the building are potentially a means by which fire could spread. The *Building Regulations 1985* require the containment of fire by compartmentalising the structure of larger buildings, and in particular multi-storey buildings (clause B3). Reference is made by the *Building Regulations* to BS 5588: 1984[3] which concerns the design of buildings in respect of the provision and preservation of escape routes for buildings in different use. Similar provisions are also made in respect of blocks of flats by CP 3, Part 1: 1971, Chapter IV.[4] The maintenance of the integrity of compartment walls, compartment floors, separating walls and protected shafts is of considerable importance in the conveyance of services throughout a building.[5] Appropriate firestopping[6] details will be required wherever any of these fire-rated structural components are penetrated by services.

REFERENCES

1 *CP 413: 1973 Ducts for building services*.
2 *Building Regulations 1985*, Part B, Approved Document B, clause B 2/3/4, Appendix B, B4.
3 *BS 5588: 1984 Fire precautions in the design and construction of buildings*.
4 CP 3, Part 1: 1971 *Flats and maisonettes (in blocks over two storeys)*, Chapter IV.
5 *Building Regulations 1985*, Part F, Approved Document F. Appendix F Pipes, clauses F8–F12, Diagram F3.
6 *Building Regulations 1985*, Approved Document B, clause B2/3/4, Appendix B, B4.

11 SERVICE EQUIPMENT FOR EXTERIOR MAINTENANCE

INTRODUCTION

The extent of maintenance to the exterior fabric of the building will depend on a number of factors including:

(i) the geometry of the design — plan shape and height; and
(ii) the nature and arrangement of exterior materials.

Design shape will to an extent dictate the ease with which access is gained to the exterior fabric, and this will also determine the amount of exterior fabric to maintain. More complicated shapes will cause the most difficulty in gaining access but probably most significant will be the influence of shape on exterior perimeter.

Figure 11.1 shows three plan shapes for a building which is to enclose 100 m² of floor area. Examination of the perimeter girths soon highlights the variation in the lengths of the walls needed to enclose the same floor area. The square building A is the most economic in terms of the wall girth while C, the most complex shape, is least economic in these terms. When the height dimension is added, the differing wall areas indicate further the significance of the design.

The wall-to-floor ratio as illustrated in figure 11.1 may be used as an indicator of the efficiency of shape to enclose a certain floor area. The efficiency of shape will of course also have significance to many of the services already included in this text: to space heating, air conditioning and refuse disposal.

If the three buildings are constructed using the same layout and types of materials, we can categorically state that there will be a greater maintenance commitment needed for shape C than shape A or B. By examination of the relationship of design to the extent of maintenance

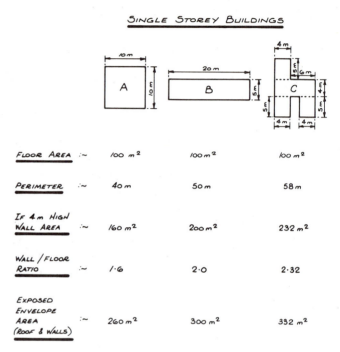

SINGLE STOREY BUILDINGS			
FLOOR AREA :~	100 m²	100 m²	100 m²
PERIMETER :~	40 m	50 m	58 m
IF 4 m HIGH WALL AREA :~	160 m²	200 m²	232 m²
WALL / FLOOR RATIO :~	1·6	2·0	2·32
EXPOSED ENVELOPE AREA (ROOF & WALLS) :~	260 m²	300 m²	332 m²

Figure 11.1   The influence of design shape on perimeter and wall area

required, a reflection of the long-term cost commitments may be assessed.

Many of the materials used on the external fabric of buildings have a low maintenance requirement but, to an extent, the demands for maintenance may be linked with the original quality of the material specification. An exception to this will be the amount of glazing provided, this requiring cleaning whether the initial outlay on this item was costly or inexpensive.

If the exterior of the building is classed into components of low maintenance needs and those of high maintenance needs, the position of these around the perimeter will have a marked effect on the type of equipment provided for cleaning and maintenance purposes.

EXTERNAL ACCESS EQUIPMENT

The equipment allowing external access to the external face of the building will largely involve access gained from ground level or access gained from roof level. Different forms of equipment vary considerably in cost and also in the rates of coverage that their deployment allows. A division may be made into permanent equipment and temporary equipment.

Temporary equipment includes:

    ladders
    stagings
    one-man systems
    adjustable platforms
    temporary cradles

Ladders obviously have limited reach and provide slow movements of the attendant operative, but it should be remembered that ladders may be used from projecting features of the building such as balconies.

The term 'staging' applies to various forms of temporary access scaffolding towers as outlined in BS 1139.[1] These, like the ladders, have a limited reach capability. Some typical details are shown in figure 11.2.

Small amounts of maintenance may be undertaken by one-man suspension systems such as the facing bicycle and the bosun's chair. These details are temporarily hung over the edge of the roof in the same manner as with the cantilever temporary cradles described later. Each of these arrangements can allow reasonable vertical movement for one attendant but horizontal movements can only be achieved by moving the suspension equipment.

Adjustable platforms are generally vehicle-mounted, hydraulically operated raised platforms of the type used for the maintenance of road lighting. This system again has limited reach but may provide quite rapid coverage of the lower regions of the building.

Temporary cradles are commonly used in cleaning and maintenance work, and are available in various sizes to suit the number of operatives needed. Temporary cradles may be suspended from roof level by either a temporary cantilever arm or from fixed anchorages.

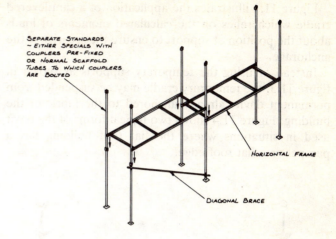

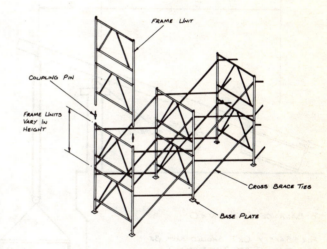

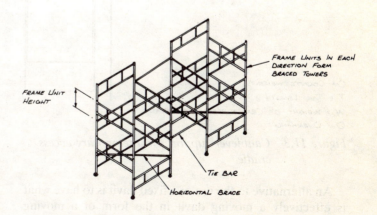

Figure 11.2 Access towers for maintenance

Figure 11.3 illustrates the application of a cantilevered cradle which relies on the calculated moments of loads about the position of support to ensure the security of the anchorage.

Instead of using the temporary supports as shown in figure 11.3, the temporary cradle may be suspended from permanent davits already anchored to the fabric of the building. Figure 11.4 shows two typical forms of the davit used in situations where the flat roof building has a parapet or a flat roof edge.

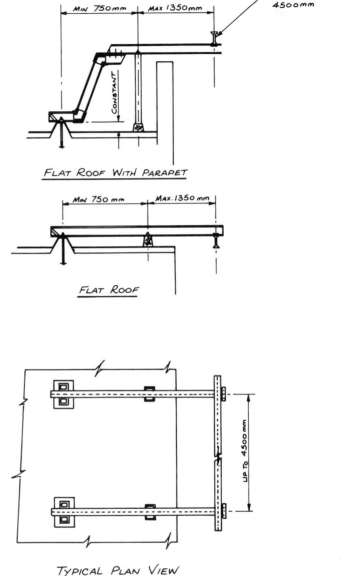

Figure 11.4   Fixed davits for the suspension of temporary cradles

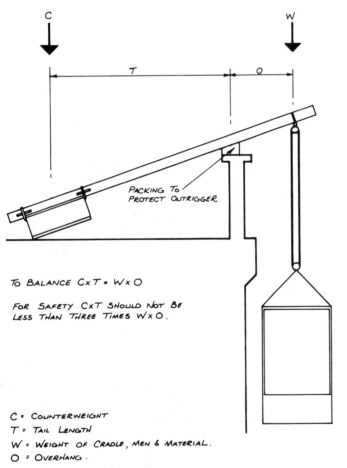

Figure 11.3   Cantilever support for a temporary access cradle

An alternative to the use of a fixed davit is to have what is effectively a moving davit in the form of a moving trolley. By having an operative located at roof level the cradle may now achieve rapid movements horizontally as well as vertically across the face of the external walls. Figure 11.5 shows examples of the movable trolley.

Cradles for temporary suspension are usually steel-framed timber details as illustrated in figure 11.6. Where cradles are to be permanently attached to suspension details such as those described later, the most common material used for the cradle body is glass reinforced

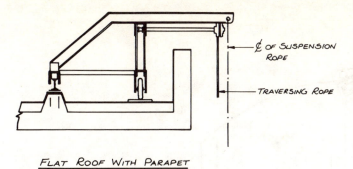

FLAT ROOF WITH PARAPET

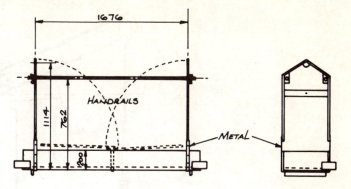

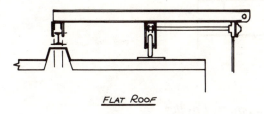

FLAT ROOF

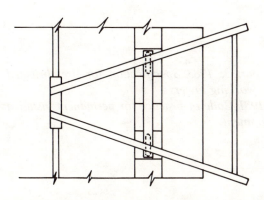

TYPICAL PLAN VIEW

Figure 11.5   Moving trolley suspensions for temporary cradles

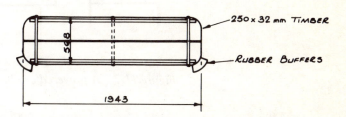

TYPICAL STANDARD TIMBER CRADLE

Figure 11.6   The steel-framed timber cradle

polyester (fibreglass) which is able to withstand the effects of weather.

Permanent forms of access equipment include:

    travelling ladders
    powered cradles

Travelling ladders are generally employed to allow access to long runs of glazing which is vertical or inclined to the horizontal. The ladders, which are made of metal alloy to resist corrosion, are mounted on a track at the head and foot of the glazing, which provides support to both ends of the wheeled ladder. Movement of the travelling ladder is achieved by the operatives pulling themselves along.

Powered cradles provide the most efficient rates of coverage of any of the external access methods. The cradle is permanently attached to a powered trolley at roof level which can be operated from a control panel in the cradle. Combinations of vertical and horizontal movements are now possible as the cradle can move vertically while the trolley is also in motion.

The powered cradle and trolley is the most expensive form of service equipment for maintaining the face of a building but it is also the form which should result in the

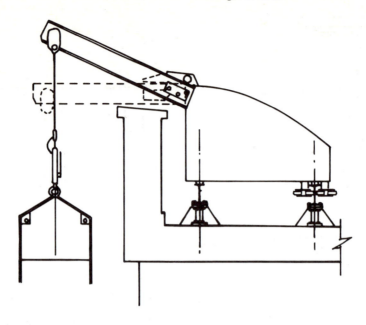

Figure 11.7   A powered cradle suspended from a powered trolley

least attendant time. It should also be appreciated that this type of access may be dictated by the design of the building if, for example, a non-loadbearing curtain walled facade is to be used. Figure 11.7 illustrates this access equipment.

BS 6037 outlines the recommendations to be followed to ensure safety in installations and operation of permanent suspended access equipment of this type.[2]

REFERENCES

1 *BS 1139: part 3: 1983 Specification for prefabricated access and working towers.*
2 *BS 6037: 1981 Code of practice for permanent installed access equipment.*

APPENDIX A: SERVICES — A BRIEF INTRODUCTION TO THE ECONOMICS OF CHOICE

The intention of this appendix is to engender thought in respect of economic questions which may be asked during assessment of the cost implications of choice options.

In the current economic climate, choice cannot simply be based on considerations of initial cost but must be made rather on comparisons of total cost — that is, initial cost plus costs in use. Costs in use of service installations would include the items:

    (i) running costs;
   (ii) maintenance costs; and
  (iii) periodic replacement costs.

Running costs relate to the consumption of fuel and in themselves may have a considerable effect on the viability of the selection process. To illustrate the effect of running cost, consider three fuel options for the space heating of a domestic property.

	Installation costs of heating system	Running costs of heating system
Electric heating	least expensive	most expensive
Solid fuel	second most expensive	second most expensive
Gas	most expensive	least expensive

This summary represents, of course, presumptions of relative costs but serves to illustrate the point.

The seesaw analogy may be applied to initial and running costs when considering cost extremes (figure A.1)

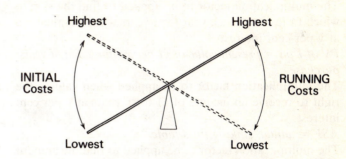

Figure A.1   *The balance between initial costs and running costs*

A similar diagrammatic representation of initial costs against maintenance costs may also be applied.

If an assessment is to be made of total costs in relation to an item of building services, then the components of total cost need to be referred to the same timebase, which is usually the time at which the comparison is taking place.

To allow the comparison of future costs, reference can be made to the use of compound interest tables. These tables consist basically of six major headings which represent lists of multiplication factors:

Amt of 1
PV of 1
Amt of 1 pa
PV of 1 pa
ASF
Annuity

Amt of 1 = amount of £1 table
The multiplication factor to be applied when finding the

sum to which £1 invested now will accumulate to in x years at y per cent interest.

PV of 1 = present value of £1 table
The multiplication factor to be applied to find the sum which should be invested now to accumulate to £1 in x years at y per cent interest.

Amt of 1 pa = amount of £1 per annum table
The multiplication factor to be applied to find the sum to which £1 invested each year for x years will accumulate to at y per cent interest.

PV of 1 pa = present value of £1 per annum table of years' purchase
The multiplication factor to be applied when valuing the right to receive an income of £1 for x years at y per cent interest.

ASF = annual sinking fund table
The multiplication factor to be applied to find the amount which should be invested annually for x years to accumulate to £1 at y per cent interest.

Annuity table
The multiplication factor to be applied to assess the annual income receivable for x years following the investment of £1 at y per cent interest.

Of these tables Amt of 1, PV of 1 and ASF may be of particular use when assessing long-term outlays to be made on alternative service installations.

For example, if inflation can be reasonably predicted the current cost of future expenditure may be assessed.

If a component of a heating system requires replacement after 15 years at a current cost of £200 and inflation is predicted at an average of 8 per cent over the next 15 years, then in 15 years' time the cost of the replacement will be:

$$\times \text{ Amt of £1 for 15 years at 8 per cent} = \frac{£200}{3.17217}$$

$$\text{Future outlay} = £634.43$$

If we know that in 20 years' time it will cost £1000 to replace a component of an air conditioning system and we

can currently obtain 8 per cent interest on investments, we can assess how much should be invested now to accumulate to £1000 in 20 years:

$$\times \text{ PV of £1 for 20 years at 8 per cent} = \frac{£1000}{0.21454}$$

$$\text{Sum to be invested now} = £214.54$$

If we would rather invest a sum annually to accumulate to the £1000 rather than place a lump sum now, the ASF table would indicate the annual saving needed:

$$\times \text{ ASF for 20 years at 8 per cent} = \frac{£1000}{0.021852}$$

$$\text{Sum to save annually} = £21.85$$

To illustrate a simple comparison of three alternative air conditioning schemes which have similar running costs but cost slightly different sums to install and have differing needs in terms of the replacement of parts while in operation:

	Initial cost (£)	Replacement costs (£)
Scheme A	15,000	3,000 every 10 years 15,000 after 20 years
Scheme B	14,000	10,000 after 15 years 25,000 after 25 years
Scheme C	14,500	4,000 after 10 years 20,000 after 20 years

Which of these schemes would cost least overall over a 35 year building life? On the face of it the answer to this question is scheme A, but evaluation of the long-term implications of the options at 8 per cent interest rate presents a much closer result than could be visualised initially:

Scheme A

Initial cost		£15,000
Replacement of parts after 10 years	= £3000	
× PV of £1 for 10 years at 8 per cent	= 0.46319	
		= £ 1389.57
Replacement of parts after 20 years	= £15,000	
× PV of £1 for 20 years at 8 per cent	= 0.21454	
		= £ 3218.10
Replacement of parts after 30 years	= £3000	
× PV of £1 for 30 years at 8 per cent	= 0.09937	
		= £ 298.11
Total cost at today's date for A		= £19,905.78

Scheme B

Initial cost		£14,000
Replacement of parts after 15 years	= £10,000	
× PV of £1 for 15 years at 8 per cent	= 0.31524	
		= £ 3152.40
Replacement of parts after 25 years	= £20,000	
× PV of £1 for 20 years at 8 per cent	= 0.14601	
		= £ 2920.20
Total cost at today's date for B		= £20,072.60

Scheme C

Initial cost		£14,500
Replacement of parts after 10 years	= £4000	
× PV of £1 for 10 years at 8 per cent	= 0.46319	
		= £ 1852.76
Replacement of parts after 20 years	= £20,000	
× PV of £1 for 20 years at 8 per cent	= 0.21454	
		= £ 4,290.80
Total cost at today's date for C		= £20,643.56

Although in this analysis scheme A is the cheapest in overall cost terms, the adoption of A must be carefully considered. Scheme B has only a fractionally larger outlay than scheme A, but initially costs £1000 less to install than does scheme A.

The value of forecasting future outlays in this way when using the compound interest formulae may be useful, but it should be viewed in its true context as it does contain weaknesses. It should be recognised that the calculations generally presume that future conditions, rates of inflation and interest rates can all be predicted. Provided the calculations are viewed in the light of these limitations, then their use may prove worthy of adoption.

Other factors, such as inconvenience or loss of production during maintenance and replacement work, would be significant features when considering the adoption of alternatives. They also show that the application of these calculations may represent an oversimplification of the situation.

APPENDIX B: THE INTERPRETATION OF DRAWN INFORMATION — SYMBOLS ASSOCIATED WITH THE COMPONENTS OF SERVICE INSTALLATIONS

These symbols are reproduced with the permission of the British Standards Institution from appropriate BS publications, including *BS 3939: Part 11: 1985 Graphical symbols for electrical power, telecommunications and electronics diagrams*, and *BS 6217: 1981 Guide to graphical symbols used on electrical equipment.*

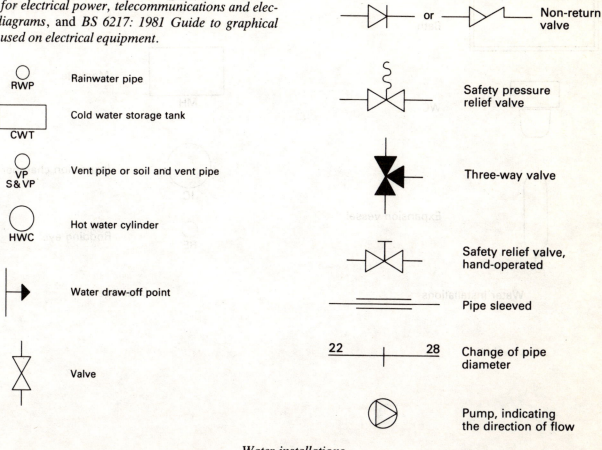

RWP   Rainwater pipe

CWT   Cold water storage tank

VP S&VP   Vent pipe or soil and vent pipe

HWC   Hot water cylinder

Water draw-off point

Valve

B   Boiler

or   Non-return valve

Safety pressure relief valve

Three-way valve

Safety relief valve, hand-operated

Pipe sleeved

22   28   Change of pipe diameter

Pump, indicating the direction of flow

Water installations

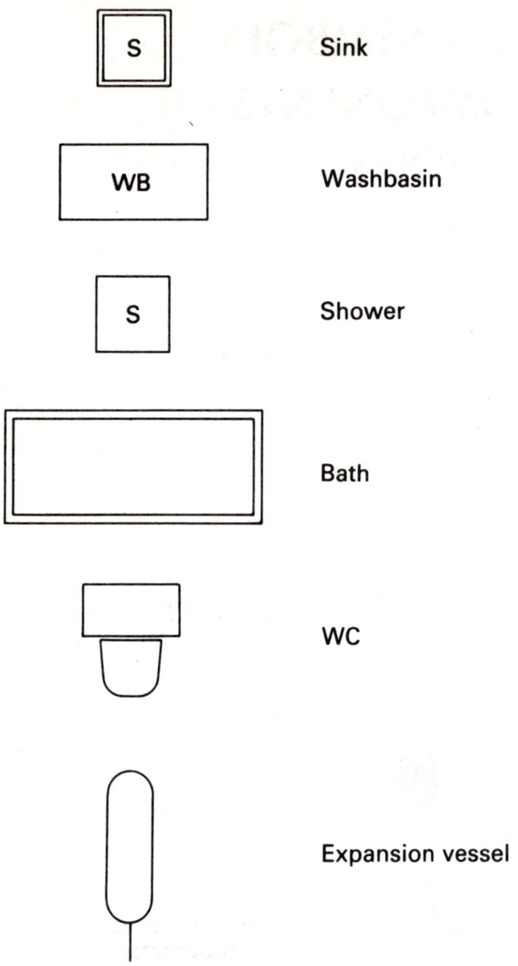

S	Sink	
WB	Washbasin	
S	Shower	
	Bath	
	WC	
	Expansion vessel	

Water Installations

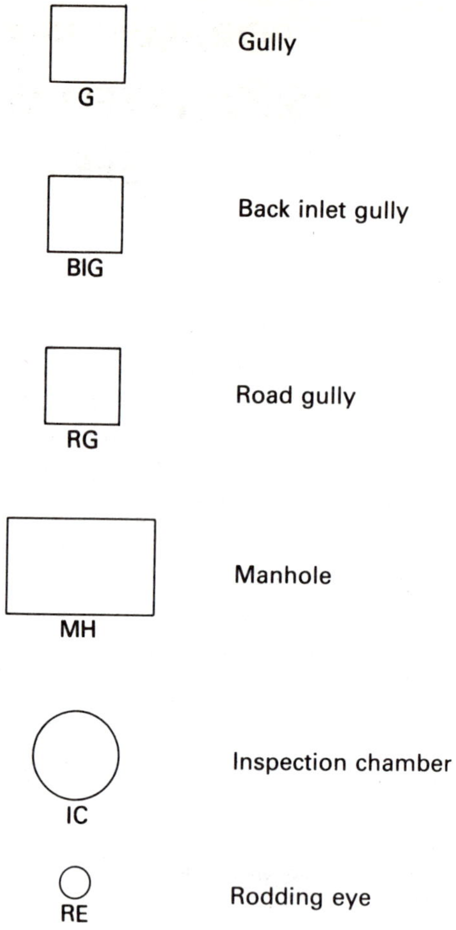

G	Gully
BIG	Back inlet gully
RG	Road gully
MH	Manhole
IC	Inspection chamber
RE	Rodding eye

Drainage

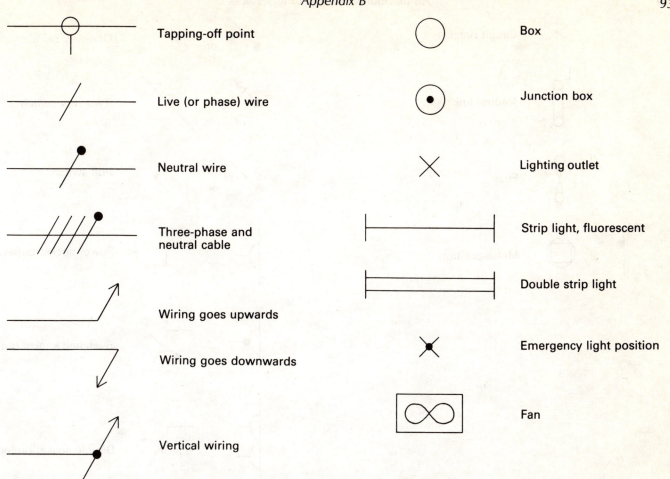

Electrical installations

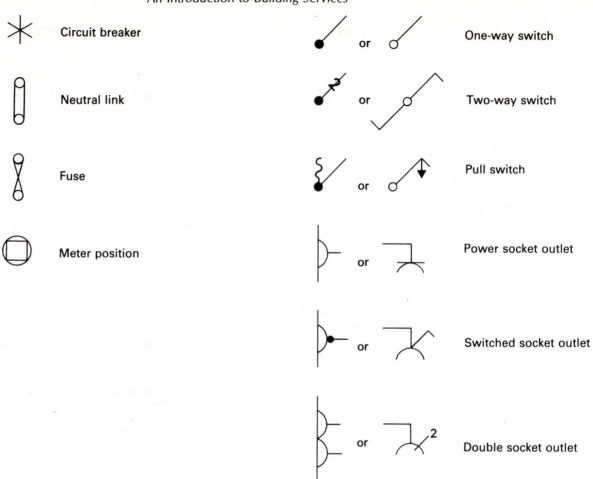

Electrical Installations

INDEX